Woburn Education Series
General Series Editor: Professor Peter Gordon

For over 20 years this series on the history, development and policy of education, under the distinguished editorship of Peter Gordon, has been evolving into a comprehensive and balanced survey of important trends in teaching and educational policy. The series is intended to reflect the changing nature of education in present-day society. The books are divided into four sections – educational policy studies, educational practice, the history of education and social history – and reflect the continuing interest in this area.

For a full series listing, please visit our website: www.woburnpress.com

Social History

The First Teenagers: The Lifestyle of Young Wage-earners in Interwar Britain
David Fowler

James Kay Shuttleworth: Journey of an Outsider
R.J.W. Selleck

Targeting Schools: Drill, Militarian and Imperialism
Alan Penn

Fairbridge: Empire and Child Migration
Geoffrey Sherrington and Chris Jeffery

Socialism and Education in Britain, 1883–1902
Kevin Manton

DICTIONARY OF BRITISH WOMEN'S ORGANISATIONS 1825–1960

Peter Gordon
and
David Doughan

With a Foreword by Sheila Rowbotham

WOBURN PRESS
LONDON • PORTLAND, OR

First published in 2001 in Great Britain by
WOBURN PRESS
Crown House, 47 Chase Side Southgate
London N14 5BP

and in the United States of America by
WOBURN PRESS
c/o ISBS
5824 N.E. Hassalo Street
Portland, Oregon 97213-3644

Website: www.woburnpress.com

British Library Cataloguing in Publication Data:

Gordon, Peter
Dictionary of British women's organisations, 1825–1960. – (The Woburn education series)
1.Women – Great Britain – Societies, etc. – Directories
I. Title II. Doughan, David
305.4´02541

ISBN 0-7130-0223-9 (cloth)
ISSN 1462-2076

Library of Congress Cataloging-in-Publication Data:

Gordon, Peter, 1927–
Dictionary of British women's organisations, 1825–1960 / Peter Gordon and David Doughan
p. cm. – (Woburn education series, ISSN 1462-2076)
Includes bibliographical references and index.
ISBN 0-7130-0223-9 (cloth);
1. Women – Great Britain – Societies and clubs – Dictionaries. I. Doughan, David. II. Title. III. Series.
HQ1945 .G67 2001
305.4´06´041 – dc21 00-012399

Printed in Great Britain by
MPG Books Ltd, Bodmin, Cornwall

CONTENTS

ILLUSTRATIONS

FOREWORD

This comprehensive dictionary of women's organisations is remarkable in its range; listings range from the Association for the Relief of Ladies in Distress through Non-Payment of Rent in Ireland, formed in 1881, to the Women's Anti-Opium Urgency Committee founded around 1908 to try and stop the opium trade from India. The well-known organisations are here, including those formed to campaign for the vote such as the Women's Social and Political Union. Also included, however, are the more obscure bodies such as various men's groups for women's suffrage and the Jewish League for Woman Suffrage.

I recognised old acquaintances from researching the history of women's political, economic and social organisations. However, while the Conservative Primrose League, the National Federation of Women Workers or the Women's League of Health and Beauty might be familiar, it is still extremely useful to have them chronicled so precisely. Many were new to me and delightful to discover. These included the Bath Club formed in 1894 to encourage exercise, particularly swimming. It was encouraging to think that women in the past had struggled to overcome laziness on chilly winter's days. The Bath Club was all the more intriguing because members also used a Morris tube rifle range. Then there was the Ladies Guild of Change Ringers, started in 1912 for both tower and handbell ringers and still going strong.

Indeed, the dictionary demonstrates the tenacious longevity of organisations. The Society of Women Journalists, for example, dates from 1893. Its house magazine, which began in 1910, was the *Woman Journalist.* In 2000, this became the *Woman Writer*. Reading the entries provides an instructive means of seeing how social attitudes shift – the University Club for Ladies transmogrifies into the University Women's Club in 1921.

Peter Gordon and David Doughan's dictionary is a valuable reminder too that so many areas of seemingly new activity have a long history. There is, for example, a cluster of associations around keeping fit; along with the Bath Club we can find the All-England Women's Hockey Association (1886) and the Ladies' Golfing Union (1893). The contemporary anxiety over wayward youth is reflected in the optimistic Snowdrop Bands of the late 1880s and the 1890s, which sought to persuade factory girls to avoid '…all wrong conversation, light and immodest conduct and the reading of bad and foolish books'. While the idea of consumer participation found an interesting expression in the 1935 Women's Gas Council, formed to enable women who used gas in their homes to '...voice their opinions and co-operate with similar organisations working for improvements in home life'. Some groups I thought might still serve a useful need. For instance, the Working Ladies' Guild, created in 1877 to assist women who were not fitted '...for the rude competition now necessary for any chance of success'. I can imagine quite a scramble to apply for such grants from academia and the media, and pleas to extend the scope to men as well as women.

The *Dictionary of British Women's Organisations* is a tribute to the organisational ingenuity of women from all walks of life in the past. Peter Gordon and David Doughan have given us a valuable reference book of impressive scholarship. I shuddered in sympathy thinking of all those hyphens they must have had to check and the inevitable confusions of 'woman' and 'women' which can arise in a compilation such as this. Yet they have contrived to produce a book which is also a pleasure simply to dip into. The *Dictionary of British Women's Organisations* not only informs, it also amuses and amazes.

Sheila Rowbotham
University of Manchester, January 2001

ACKNOWLEDGEMENTS

The authors would like to thank the following organisations for permission to reproduce illustrations: the Women's Engineering Society for illustrations 1 and 2; the Co-operative Women's Guild and *Co-operative News* for illustration 8; the Women's Royal Voluntary Service for illustrations 9 and 10; the National Council of Women for illustration 12; and the National Federation of Women's Institutes for illustration 19. If any other persons or organisations believe they have claim for reproduction rights to any other illustrations, we ask them to contact the publisher.

INTRODUCTION

Organisations of all kinds tend to be taken for granted, and it is only occasionally that there is an impulse to look closely at what they are and what they do. For example, in the late 1980s and early 1990s some of us in Britain were intrigued to note the interest shown by visitors from Eastern and Central Europe in the workings of our associations, clubs, unions and societies; we would reluctantly turn up out of a pure sense of duty at excruciatingly boring annual general meetings and other business sessions to find our visitors eagerly taking notes and asking complex questions about procedure and the way in which these bodies were organised. This was, of course, due to the concern with civil society in its various definitions from Locke through Leo XIII to Havel, for one of the most crucial features of civil society is the freedom to organise autonomously, independently of church, state or police. In Britain, and especially England, we pride ourselves on the fact that we have never needed to get official approval if we want to start up a philatelic society or a snooker club, unlike our less fortunate continental brethren. This is part of the inalienable heritage of an Englishman.

Historically, however, it has been much less likely to be part of the heritage of an Englishwoman, Scotswoman, Irishwoman or Welshwoman, who, as in the much-lauded Cleisthenic democracy of ancient Athens, was not considered to form part of civil society, or the polity in general. Both élite men and working men have had their clubs, parliaments, unions, lodges, guilds, congregations and the like, which may have been more or less inclusive, but the one category that was traditionally excluded from most, if not all, of them was that of women. In what came to be expressed as the separate spheres ideology of the nineteenth century, women belonged in the private sphere, while organisations were emphatically public. In practice this did not prevent women joining some 'male' organisations, or forming women's sections or affiliates of men's organisations, or even on rare occasions organising independently (especially for the abolition of slavery), but it was not until the early to mid-nineteenth century that they started forming a large range of organisations. In fact there had been previously, and continued to be, many bodies working for the welfare, real or supposed, of groups of women (governesses, widows, emigrants, prostitutes, etc.), but their organisation had been mainly or entirely in the hands of men; women's autonomous organisation was relatively unprecedented.

Of course, there had been informal networks of women of all social stations, as of men, time out of mind, but to come together in a group organised around a particular issue (even if that issue was entertainment or needlework) was to replace the informal and covert with the formal and overt. Even to form a ladies' social club was a mild assertion of women's right to public space. The reasons why women's autonomous organisations start to appear in such numbers at this time seem to be complex, and they probably have much in common with the reasons for the growth of a conscious women's movement, as well as for general movements for a greater inclusivity in public life. In fact, the women's movement itself both gave rise to a considerable number of women's organisations and was to some extent dependent on them for its own development. For these reasons, in compiling this dictionary we have taken 1825 as a convenient, if somewhat arbitrary, starting point.

Women-specific organisations were established largely as a response to women's exclusion from the organisations long since developed by men. However, in forming their organisations, they relied to a very large extent on male models of procedure, organisation and internal relationships. Annual general meetings and annual reports, with audited balance sheets, abounded. Meetings were presided over by a chairman (*sic*), to whom remarks were addressed, and officers were elected by the membership in accordance with a set of rules, or even a constitution. Some of these bodies became limited companies, or trusts, or charitable bodies, and were thus brought within the ambit of the law. Procedural manuals were studied intensely. The 'male' structures were taken as given, and were hardly questioned until the 1960s. 'Structurelessness' does not seem to have even been considered; if it had been, it would probably have been perceived as a retreat to the covert 'feminine' stance from which these organised women felt themselves to be advancing. The established procedures were perceived to work, and there was no particular desire to go beyond them. Indeed, a major reason for this adoption of 'masculine' styles was to demonstrate that women could make them work as well as men could. In terms of organisation at least, notions of equality prevailed over difference, even in those schools and colleges which were expressly rejecting the straitjacket of the traditional male syllabus. Girls' schools and women's colleges ran debating, literary and social societies along the same lines as their male counterparts.

These are, by and large, the sort of organisations with which this dictionary is concerned, and which flourished almost to the exclusion of any alternatives in the period 1825–1960; indeed, despite the questionings and experiments of the past four decades, they remain typical. However, this is one of the few things they have in common. They vary enormously in their aims, their methods, their relation to each other and to other organisations, to the state, to the economic and social order and in their attitude to femininity and to feminism. There are purely social clubs, there are reforming clubs, professional clubs, political clubs; organisations for and against women's participation in public life, in parliament, in the churches; organisations asserting women's right to take part in a whole range of sports; organisations for and against various conflicting ideals of sexual reform, or new ethical and spiritual concepts; organisations promoting women's equality in the workplace side by side with organisations of women in particular areas of work devoted at least partly to the promotion of mutual interests; the list is far from exhaustive.

The variety and lack of common purpose of these organisations is a function of the fact that women are people, and are as heterogeneous as men. All they have ever had in common is their subordinate status, and the fact that for women to organise independently at all could be seen as challenging, however obliquely, the established order. Indeed, there has even occasionally been the apparent contradiction of anti-feminist women organising in order to deplore the organisation of women, and campaigning in public against women appearing in the public sphere (this is most blatant in the case of the Women's National Anti-Suffrage League). Some of these organisations have attracted far more scholarly attention than others, but the interest expressed in certain bodies has depended less on their size, or their public visibility, than on how they fit an accepted academic paradigm, usually of bodies involved in explicitly political campaigns. Thus a great deal has been written on the Six Point Group, which did indeed have political influence, as well as having several prominent authors among its members, but was very small, with a membership usually numbering less than 300. By contrast, disproportionately little research has been done on the National Federation of Women's Institutes, whose membership was nearly 1,000 times larger than that of the Six Point Group at the time of the latter's greatest appeal. Another factor that attracts attention is the internal politics of an organisation, where a relatively placid body such as the National Federation of Women's Institutes provides less graphically complex material for researchers into the dynamics of organisations than, say, the Women's Social and Political Union, with its notorious four major splits in eight years. To cite an even more extreme example: the Council of Married Women's membership numbers make even the Six Point Group look substantial, but its members were so prominent, and it generated internal politics to such an extent, that merely to give a basic account of them takes a disproportionate amount of space.

A great deal of work remains to be done on the individual members of most of these organisations, the relatively silent minority. Indeed, even the prominent office-holders have barely been touched on by researchers. Assertions and assumptions are often made about networks of women. There were indeed such networks, and some of the women mentioned in this dictionary were visibly active in more than one organisation. A glance at the index of this work will give some idea of the breadth of activity of such differing women as 'Mentia' Taylor, Louisa Hubbard, Maude Royden, Emmeline Pethick-Lawrence and Lady Denman. However, a further assumption is sometimes made, that the activists in these organisations mainly belonged to a relatively small circle, and possibly spent much time and energy taking in each other's washing. A study of these organisations reveals the extent to which this impression is misleading. The number of women who held office in several organisations is much lower than the number of those who appear only once in these entries, and about whom it would be difficult in the extreme to find personal information. Certainly the amount of overlap among office-holders (and possibly among members) is much less than is sometimes assumed. When there is significant overlap, it tends mainly to be among organisations with a similar, if not identical aim, the most notable example being that of the suffrage societies. Outside this, it can scarcely be said that women's organisations have largely been run by a handful of activists all known to each other and unrepresentative of women as a whole.

However, there is in fact one way in which the women who held office in these organisations were not typical of the majority of women. In order to fulfil the obligations of office they needed money, and even more, time – freedom from the exhausting demands of full-time work or running a household. A sympathetic husband or supportive parents were a great help, but best of all was some sort of independent income (even a pension or annuity) which would permit women to take on the business of arranging meetings, collecting subscriptions, the necessary level of which could be another factor limiting ordinary women's participation, or even stuffing envelopes, as Virginia Woolf did for the People's Suffrage Federation. It is difficult in most cases to ascertain the precise financial circumstances of the office-holders of these bodies; however, it is noticeable that a disproportionate number of them are titled ladies. In some cases such women may well have been figureheads put in positions that were effectively non-executive merely in order to give the organisation an air of respectability. Nonetheless, these are likely to be precisely the sort of women who had the time and resources to devote to running organisations, which was not the case, for example, with women Chartists, whose organisations were correspondingly obscure and apparently short-lived. Some bodies became aware of this imbalance, and tried to take measures to redress it. In a few cases, subscriptions would be raised to meet the expenses of a particular member whose work was especially valued; more constructively, in the years leading up to the First World War, the relatively wealthy National Union of Women's Suffrage Societies employed working-class women such as Selina Cooper and Sarah Reddish as 'organisers', thus enabling them to work full-time campaigning for women's suffrage. Such cases were, however, rare exceptions.

Financial considerations also influenced the enormous variation in organisations' premises. Probably none could match the Lyceum Club in the 1920s, with its Piccadilly address, its ballroom and interior decoration by the Adam Brothers and Angelica Kaufmann, though even organisations which would not have described themselves as social clubs sometimes had more modest versions of this for their members' use; for example, the relatively small and highly political London and National Society for Women's Service in the 1930s had premises with a library and a lecture hall in the heart of Westminster. However, it was more usual for an organisation to have simply an administrative office, or even to be run from the home of the secretary or some other private address, with rooms being hired for meetings. This of course depended to a large extent on the wealth or otherwise and the size of the membership of the organisation, and the rate at which subscriptions were set. An inconsistent attempt has been made here to indicate this, and addresses, when the information has been found.

As has already been indicated, assumptions based on recent evidence are frequently made about individual organisations. For example, the Women's Institutes are often written off as a middle-class women's organisation devoted to making jam and singing *Jerusalem*, a very misleading impression, for which admittedly the Institutes themselves bear some responsibility. In another instance, the fact that the Women's League of Health and Beauty's original motto was 'Racial Health' has led some commentators to assume that it formed a sort of British equivalent of the Bund Deutscher Mädel, or at least had racist sympathies, which is to ignore both the usage of the time and the League's own history. There is also the large category of women's clubs, which are often lumped together in order to be dismissed, when in fact they represent a diversity of subcategories, without even considering individual differences. For instance, there is the case of clubs designed for young working-class women, which have only been the subject of research when they are connected with themes or individuals that fall into more generally accepted areas of research; a good example is the Espérance Girls' Club, with its links to women's suffrage on the one hand and the folk-dance revival on the other. Even among more conventionally social clubs there is a great deal of work remaining to be done. In their way, the Pioneer and Lyceum Clubs were possibly as important for prominent women in the fields of arts and literature as many of the better known men's clubs of the day, for example, the Savile, the Garrick, the Savage, the Travellers', the Thatched House, etc., as well as stimulating women's self-improvement and general intellectual awareness. Surprisingly, they appear to have attracted little or no serious research.

The fact that some bodies are under-researched may be at least partly due to the level of difficulty encountered in seeking information on some societies as compared to others. While there is no problem finding exhaustive details of, say, the major women's suffrage societies or the Women's Institutes, information on other bodies, major and minor, is often scattered, partial (in both senses), or so exiguous as to be almost non-existent. An example is the Women's Printing Society, an extremely important venture which involved two women very prominent in the history of British feminism, Emma Paterson and Emily Faithfull, but on which information is at best scrappy and usually unhelpful, throwing no light, for instance, on why the *Women's Penny Paper* stopped using the Women's Printing Society in 1890, opting instead for a more conventional male printer, and thus dropping the proud boast 'the only paper … printed by women' from its masthead.

The role of men in women-oriented organisations is both interesting and complex. Men may appear as simple supporters, members of organisations run by women for women; or they may organise separately as men supporting women, in fact or in theory. The latter group may be problematical, especially when 'protecting' prostitutes or other 'fallen' women, or women deemed to be at moral risk. Even when men organise to support such obviously feminist campaigns as women's suffrage, their contribution may be less straightforward than appears on the surface (for further discussion of this theme, see John and Eustance, 1997). All we can do in the case of an organisation run by men, or in which men seem to take a prominent part, is to suspend judgement and investigate further.

Although women's organisations were so often structured along 'male' lines, the fact of women-specific orientation often inspired uneasiness, or even hostility, among male-dominated or 'mixed' bodies, not least among their women members. The 1920s Labour Party women's conference resolutions urging women Party members not to join single-sex organisations because of the risk that 'they would be in danger of getting their political opinions muddled' were unusual only for their plain speaking. In another instance, the historian of the Federation of Women Zionists likens its relationship with the general Zionist Federation to a marriage facing constant conflicts, or, interestingly, a divorce that never quite becomes absolute.

SCOPE

As stated above, the scope of this dictionary is British women's organisations between 1825 and 1960 (including Irish women up to 1921, but excluding international organisations). It includes as many non-commercial organisations as the authors can identify that were wholly or principally run explicitly by women, usually for women. The fact of women holding high office in a 'mixed' organisation does not necessarily qualify it; even such an extraordinarily 'mixed' group as the Men and Women's Club has not been felt to fall within the purview of this work, mainly because it was principally the creation of Karl Pearson. On the other hand, a number of organisations have been included for completeness that were not necessarily run wholly by women, but which aimed to support women's campaigns, such as the Abortion Law Reform Association, the National Association for the Repeal of the Contagious Diseases Acts, dress reform associations or men's organisations in support of women's suffrage. Again, the selection has been limited to formally established organisations: networks like the Langham Place Circle, however influential, have been excluded, partly because these 'circles' frequently overlapped with others, but mainly because they are too amorphous to identify clearly (although an exception has been made in the case of women inspectors, where the informal networks were acting so much like formal organisations, and to specifically women-centered ends, that their exclusion would have been invidious).

Also, specifically business organisations run by women for purely commercial ends have been excluded, as have explicitly governmental bodies, such as the Women's Advisory Committee of Reconstruction, as not being truly autonomous, although here again there is a grey area; and we have included some quasi-autonomous bodies which nonetheless had strong links with governmental departments, such as the Women's Advisory Council on Solid Fuel, or Queen Alexandra's Royal Army Nursing Corps. Similarly, women's organisations within political parties have been given the benefit of the doubt, despite the fact that their autonomy has often been severely qualified and, as in the case of the Primrose League, they may not even be women's organisations in the strict sense. The entries are intended to describe the bodies purely from their organisational characteristics, and there is no attempt to give an outline account of the campaigns in which they were involved.

Girls' schools have been excluded from separate entries mainly because of their sheer number and similarity, but also because they are usually affiliated to such national organisations as the Girls' Public Day Schools Trust. Similarly, women's colleges within universities do not have separate entries, although a glance at the index will reveal that most of them are mentioned in entries for national organisations or settlements. However, some attempt has been made to identify the founders, leading members and addresses of organisations, or at least their locations (though see above on the limitations of this attempt), together with the body's periodical organ, where one existed. When an address is given, it is in the form stated in the original source. No attempt has been made to reconcile addresses with post codes current at the time of writing.

A metropolitan bias will be perceptible in many of these entries, which reflects the tendency of most of these bodies to have headquarters in London. It also reflects the availability of sources, as well as the authors' own bases. However, there may be more local and regional organisations in England, Wales, Scotland and especially pre-1921 Ireland which deserve inclusion, but have been omitted purely because of the authors' ignorance of them.

A few entries of bodies which do not fulfil any of the above criteria have been included, effectively to warn researchers against wasting time pursuing misleading names, such as the Women's Imperial Defence Council. They led us astray; we do not see why they should do the same to anybody else.

FORMS OF EXPRESSION

The terms chosen to describe office-holders, and so forth, have been those used by the organisations themselves, and not brought into line with current preferences – for example, the term 'chairman' is almost universally employed, hence reproduced in the entries. Personal names have also been given within the entries in the form that they usually appear in the organisation's own documentation; whenever possible, in the index to entries they have been given their full correct forms, occasionally together with variant forms that may be met, e.g. 'Strachey, Rachel Conn Costelloe (Ray)(*Mrs* Oliver)'; the same courtesy has not invariably been extended to the men who appear in the index. However, in a surprising number of cases, the authors have so far been unable to ascertain many individual women's own given names, and they have reluctantly been made to appear either with initials only, or simply as 'Miss' or 'Mrs', or worse, under their husband's name (e.g. Mrs Alfred Watts). It should be noted that the titles 'Miss' and 'Mrs' have only been used in cases where no further information was available. In other instances no indication of marital status is given.

SOURCES EXAMINED

These have been extremely various, ranging from such compilations as the *Englishwoman's Year Book*, through the periodical press (both general women-oriented weeklies and monthlies, such as the *Queen,* or *Shafts*, and organs of the organisation under consideration, indicated as such in the entries), to a wide variety of specific organisational material, such as published annual reports, yearbooks, handbooks and unpublished archives held in a number of locations, most prominently The Fawcett Library and the British Library (St Pancras and Colindale), but also the British Library of Political and Economic Science, Dr Williams's Library, the Imperial War Museum, the Institute of Education, University of London, the Jewish Studies Library at University College, London (formerly the Mocatta Library), the London Library, the Royal College of Nursing, the University of North London, and the Wellcome Library for the History of Medicine. Our thanks go to the staff of all the above, with special recognition of the assistance (and forbearance) of David Doughan's colleagues at The Fawcett Library, and particular acknowledgement of the contribution of Virginia Clark.

ARRANGEMENT

This is alphabetical, according to the usually accepted name of the organisation, with cross-references from variant names, or successor / predecessor organisations. References within the text to other organisations with an independent entry ('see' references) are indicated in bold. At the end of an entry, there is frequently a reference to a work mentioned in the Bibliography considered useful for further reading, though it is not intended to be a comprehensive bibliography of sources. These references are set out according to the Harvard reference system, that is, author and date, with full publication details in the Bibliography. Finally, there is an index of names, places and subjects at the end of the book.

This work is the first attempt to bring together information on this large and heterogeneous group of organisations, with the authors adapting and changing their methodology (and parameters) as the work progressed, and as such it suffers from various inconsistencies, for which the authors beg the readers' indulgence. It is also certain that many organisations have been excluded for no reason other than that the authors have overlooked them, or have been unaware of their existence. The authors would be very interested to know of any such, together with details of any sources of information on them, for a possible future edition.

Abortion Law Reform Association

This was formed in 1936. By the mid-1930s, reports of maternal mortality had revealed the existence of an illegal culture of back-street abortion, and for this among other reasons growing numbers of contraception advocates came also to support the decriminalisation of abortion, illegal under the 1861 Offences Against the Person Act, only marginally modified by the Infant Life (Preservation) Act of 1929 – although Marie Stopes remained resolutely opposed. A major turning point came when the **Women's Co-operative Guild** approved reform of the law in 1934, and finally the Abortion Law Reform Association was formed, with Janet Chance as chairman and Alice Jenkins as honorary secretary. Its initial objects were stated to be the ending of a pregnancy on request where necessary to preserve the woman's physical or mental health, or when there was a risk of birth defects, or when pregnancy resulted from rape. Its campaigning remained relatively low key until the early 1960s, when Vera Houghton was chairman, and committee members included Dr Martin Cole and Madeleine Simms, who edited the Association's newsletter. Under the Association's auspices a parliamentary campaign was undertaken which resulted in David Steele's Private Member's Bill becoming the Abortion Act of 1967. At the time of writing, the Association continues to campaign for women's reproductive rights.

(Leathard, 1980)

Active Service League

Started in February 1914, this was a suffrage ginger-group under the auspices of the **National Union of Women's Suffrage Societies** with the intention of broadening the organisation's base, especially by attracting younger women. It was mainly based in Oxford. It did not survive the outbreak of the First World War.

(Crawford, 1999)

Actresses' Franchise League

Founded in 1908 by Gertrude Elliott, Winifred Mayo, Sime Seruya and Adeline Bourne, this represented actresses of both militant and non-militant suffrage tendencies, and included a number of famous names, such as Lena Ashwell, Edy Craig, Cicely Hamilton, Lillie Langtry, Kitty Marion, Elizabeth Robins, Athene Seyler and Sybil Thorndike. Its main activity was producing suffrage propaganda plays in conjunction with the **Women Writers' Suffrage League**. Its colours were pink and green. During the First World War it travelled round military camps and hospitals entertaining and propagandising in good agitprop style. It remained in existence until at least 1934.

(A. J. R., 1913; Crawford, 1999; Stowell, 1992)

Albemarle Club

Established at 13 Albemarle Street, London,W1, in 1881 by a progressive group including Elizabeth Malleson and Henry Holliday, the club was open to both men and women. Prominent members included Millicent Fawcett, Lilias Ashworth Hallett and May Sinclair. It moved later to nearby premises at 37 Dover Street, and was still active in 1923.

Alexandra Club (1)

Founded in 1884 at 12 Grosvenor Street, Piccadilly, this exclusive club was for ladies eligible to attend the Queen's Drawing Rooms, which seems to have excluded (among many others) divorcees. Residential accommodation was provided. One of its rules forbade the entertaining of male visitors on the premises and smoking was not allowed. The entrance fee was five guineas, and annual subscriptions were five guineas for town members and four guineas for country members. It was still in existence in 1928.

Alexandra Club (2)

A club at 133–5 St Stephen's Green, Dublin, established in 1892 for social purposes. Membership was by nomination of two members and election by the Committee. Accommodation was provided on a permanent or temporary basis. In 1906 there were 350 members. The entrance fee was two guineas, as was the annual subscription. The secretary was Miss Ewing Kennedy. It was affiliated to a number of London Clubs including the **Green Park Club,** the **New Victorian Club**, the **Grosvenor Crescent Club** and the **New Era Club**.

All-England Women's Hockey Association

Although the all-male Hockey Association was formed in 1886, it was not considered a major sport at public schools. The first private women's hockey club was founded the following year. Following the lead of the Irish Ladies' Hockey Union formed in 1892, a meeting was held in November 1895 at Westminster Town Hall of representatives of women's colleges from Oxford, Cambridge and London as private ladies' clubs. The Ladies' Hockey Association was subsequently established with Miss Lilian Faithfull, head of the Ladies' Department, King's College, London as its president. After a rebuff from the Hockey Association for affiliation, the LHA set up the All-England Women's Hockey Association (AEWHA), stipulating that no man could ever hold an executive office in that Association. The first county team was formed in Sussex in 1898, to be followed by many others. Welsh and Scottish Associations were established in 1902 as was the official organ of the Association, the

Hockey Field, issued as a weekly paper. In the same year, four territorial organisations – North, South, West and Midlands – were formed in England to administer the game. The sport became popular at school and college levels and the AEWHA was involved in making rules, fixing fees and selecting an England team. The measure of its success can be seen in the spread of the sport by the outbreak of the First World War to include five territorial and 36 county associations, more than 300 school, college and private clubs and matches at all levels including the international. By 1939, there were over 2,000 affiliated clubs.

(All-England Women's Hockey Association, 1950; McCrone, 1988)

Alliance Club

Advertised as 'for ladies interested in Christian work', the Club had an Evangelical basis. It was situated at 61 Curzon Street, Mayfair, London, W. The entrance fee in 1906 was two guineas with annual subscriptions varying between two and three guineas.

Amazon Defence Corps

Soon after the inauguration of the Local Defence Volunteers (later the Home Guard) in June 1940, women started campaigning to become members. This corps appears to have been one of the earliest organisations attempting to have women helping to defend Britain. Its aims were to obtain training for women in the use and handling of firearms and other weapons of defence; that trained women should be admitted as members of Local Defence Volunteers on terms of equality with men; to urge training in the use of firearms for all women of the services; and to encourage in women the spirit to resist the invader by all means available. Membership was one shilling per annum. The organising committee included Venetia Foster, Nora Robbins, L.E. Alton, Juanita Francis, Elizabeth MacDonald and Edith How-Martyn. Its activities may have been continued by **Women's Home Defence**.

Anglican Group for the Ordination of Women to the Historic Ministry of the Church

The Group was formed in about 1931 in the wake of the 1930 Lambeth Conference, although its first annual general meeting did not take place until 1933. Its aim was to secure effective recognition of the Christian principle of spiritual equality between men and women by urging upon the Church of England the admission of women to the threefold ministry of bishops, priests and deacons. Its vice-presidents were the Bishop of Lichfield and the Dean of Canterbury; its Council included several prominent Anglican clergymen as well as prominent women, such as principals of women's colleges, the Headmistress of Roedean School and the actress Dame Sybil Thorndike; its first secretary was Mrs Ronald Best. It was wound up in the mid-1970s.

Arachne Club

This Club, founded in 1905 at 66 Russell Square, London,W, was for training women in cooking and general domestic work. Its secretary was Mrs Armstrong.

Artists' Suffrage League

The League was founded in January 1907. Its chairman was Mary Lowndes, and its members were practising artists, including the Australian, Dora Meeson Coates, Mary Sargeant Florence, Bertha Newcombe and Caroline Watts, designer of the 'Bugler Girl'. Its main activity was producing visual suffrage propaganda in the form of cartoons, often in conjunction with the **Women Writers' Suffrage League**, postcards, posters and especially banners (Mary Lowndes's design book for posters and banners is held in the Women's Library). It worked mainly for the **National Union of Women's Suffrage Societies**; its artistic techniques were relatively conventional as compared to those of the **Suffrage Atelier**.

(Crawford,1999; Tickner, 1987)

Associate Institution for Improving and Enforcing the Laws for the Protection of Women

Established in 1844 by the amalgamation of a number of societies which had sprung up during the late 1830s and 1840s, to promote female protection. It sought to legislate against brothels and those who profiteered from the earnings of 'fallen women'. Lord Robert Grosvenor was chairman. It was apparently male-dominated. Its organ was the *Female Friend.*

(Gleadle, 1995)

Association for Moral and Social Hygiene

This Association was formed in 1915 from an amalgamation of the **Ladies' National Association** and the British section of the British, Continental and General Federation for the Abolition of Government Regulation of Prostitution. Its first, and extremely influential, secretary was Alison Neilans, who held the post till her death in 1942. The Association continued the tradition of its predecessors in opposing state regulation of prostitution, notably in regard to the provisions of the Defence of the Realm Acts in both World Wars (Sections 40D and 33B respectively). Additionally it was concerned with a wide range of issues relating to sexuality in general, especially sex education. It took a more consistently feminist line than other contemporary organisations, such as the **National Vigilance Association**. Unlike many other progressive organisations between the wars, it had strong reservations about eugenic policies, and was particularly opposed to measures allowing the sterilisation of the unfit. After the Second World War it seems to have lost some of its campaigning vigour; for example, it was not particularly assiduous in making representations to the Wolfenden Committee on Homosexual Offences and Prostitution, 1954–57, though it became very vocal in its criticism of the 1959 Sexual Offences Act, which in part emanated from the report of that Committee. In 1953 it changed its name to the Josephine Butler Society. Its organ was the *Shield.*

Association for Promoting the Higher Education of Women in Cambridge

The creation by Cambridge University of the Higher Local Examinations for women over 18 years of age in 1869 led to a group of Cambridge individuals, led by Henry Sidgwick, the philosopher, and Millicent Garrett Fawcett, to press for lecture courses to prepare women living within 80 miles of the city for the examination. As a result, a lecture committee was formed which became the Association for Promoting the Higher Education of Women in Cambridge. The enterprise was a successful one but the Association was unable to bear the responsibility of further expense towards women's education. Within two years, a residence for students had been acquired at Cambridge and Sidgwick persuaded Anne Jemima Clough, a leading figure in the **North of England Council for Promoting the Higher Education of Women**, to take charge. By 1872, on Sidgwick's initiative, a new residence, Newnham Hall, had been built. Clough became its first Principal with its staff of tutors.

(Burstyn, 1980; Kamm, 1965)

Association for Promoting Trained Nursing in Workhouse Infirmaries

Although Florence Nightingale had attracted many middle-class women to became trained nurses, the Poor Law infirmaries presented a different picture. Poor Law Guardians were reluctant to pay the same rates as voluntary hospitals or to provide similar standards of accommodation and working conditions. As a result, much of the nursing in workhouses in the nineteenth century was undertaken by pauper women selected by the matron, in some cases in return for beer and 2s 6d a week. Despite protests by the Poor Law Board on the employment of pauper nurses, no official action was taken. A number of ladies had undertaken the task of helping to supply Boards of Guardians with trained nurses for workhouse infirmaries, especially in London, Liverpool and Birmingham. Miss Louisa Twining had been instrumental in setting up the **Workhouse Visiting Society** in 1861 and she was determined to bring about workhouse reform. After reading a report in the *Lancet* in 1866 outlining the hardship and neglect of the poor sick in workhouses, she campaigned for the training and supply of nurses and matrons for these institutions. This led to the establishment of the Association for Promoting Trained Nursing in Workhouse Infirmaries in 1878 'to promote employment of paid and efficiently trained nurses in all workhouse infirmaries and sick asylums', and she became its first secretary. Twenty years later, there were 800 nurses employed in workhouse infirmaries. One of her triumphs was the order of the Poor Law Board in 1897 forbidding the employment of pauper nurses, though they continued to work in the infirmaries under the supervision of a trained nurse. A superintendent nurse of at least three years' standing was to be employed in all infirmaries with a staff of three or more nurses.

(Crowther, 1982; Twining, 1898)

Association for the Aid and Benefit of Dressmakers and Milliners

This body was established in the mid-nineteenth century to assist the workers in these exploited trades and act 'as a noble breakwater against the inroads of oppression and a valuable countervailing agency to the force of temptation'.

(Crow, 1971)

Association for the Extension of the Contagious Diseases Acts

The Association was formed in 1869 to campaign for the extension of the provisions of the Contagious Diseases Acts (1864–69) to the whole of the United Kingdom (they had previously been restricted to specific named ports and garrison towns). This would have entailed national regulation of prostitution on a model similar to that practised at the time by the French state. Its activity appears to have ceased with the repeal of the Acts in 1886.

Association for the Relief of Ladies in Distress Through Non-Payment of Rent in Ireland

Formed in 1881 to render financial help to ladies whose income was affected by the Land War in Ireland and to raise money for this purpose. Two local committees were also set up in London. Though the management was mixed, women only visited those requesting help. In its first two years, 1,076 grants or loans were made.

(Luddy, 1995)

Association of Advertising Women

This organisation was set up in 1910 to bring together, for mutual aid and social enjoyment, women workers actively engaged in advertising work. By this time there were many employed as copywriters, artists and designers and 'canvassers' selling advertising space. In 1916 its headquarters were at 154 Clerkenwell Road, London, and its president was Miss Ethel M. Sayer. Subscriptions were two guineas and one guinea according to status. It had disappeared by the end of First World War.

Association of Assistant Mistresses

Established in 1884 as the Association of Assistant Mistresses in Public Secondary Schools to safeguard the interests of women teachers in secondary schools on matters such as salaries and conditions of service, and to provide a forum for the discussion of educational issues. Its formation was largely due to the support of the **Association of Head Mistresses**; its first annual general meeting was held at the North London Collegiate School in 1885, where Frances Buss was headmistress. The Association expanded after the 1902 Education Act increased the provision of secondary education for girls. It raised several matters of importance for the profession: for example, at its annual general meeting in 1907, the Association expressed its opposition to the uniformity of timetables insisted upon by the Board of Education and urged the recruitment of more women inspectors to the central Inspectorate. The majority of its members were graduates who taught in grammar schools. It changed its name to the Association of Assistant Mistresses in 1922. Increasingly thereafter the Association co-operated in negotiations over pay and working conditions with the Incorporated Association of Assistant Masters (IAAM), the Association of Head Masters, and the Association of Head Mistresses; in this connection the four groups were known as the Joint Four. In 1978, the AAM formed an amalgamation with the IAAM to become the Assistant Masters' and Mistresses' Association (AMMA).

(Gosden, 1972)

Association of German Governesses in England

This Association was established in 1876 under Royal patronage at 16 Wyndham Place, Bryanston Square, London. It admitted only German ladies who showed evidence of fitness for the post of governess. It had a Gentlemen's Committee and a separate Ladies' Committee, presided over by Lady Suffield. Between 1870 and 1887, 2,344 English families and schools were provided with governesses in connection with the Association.

Association of Head Mistresses

Founded in 1874 by Frances Buss, Dorothea Beale and seven other secondary school headmistresses 'in order to know what we ought to assert and what to surrender'. Its original title was the Association of Head Mistresses of Endowed and Proprietary Schools but as different types of secondary schools developed, membership was open to the heads of all such schools. Its main objectives were to advance girls' secondary education and to discuss matters of mutual interest at annual conferences. In 1976, it became absorbed into the Secondary Heads Association.

(Glenday and Price, 1974)

Association of Hospital Matrons

This Association was formed 1919 when the Nurses' Registration Bill for state registration allocated a number of seats on the proposed General Nursing Council to the **Matrons' Council of Great Britain and Ireland**. Three matrons, Miss Musson, Miss Lloyd of St Thomas's Hospital and Miss Cox-Davies of the Royal Free Hospital, wrote to 50 matrons asking for support for a 'really representative association of hospital matrons'. At a meeting in May 1919, the setting up of an Association was agreed to by the 89 matrons present. Two months later, it had a membership of 183.

Association of Infant Consultations and Schools for Mothers

A department of the National League for Physical Education which was established in 1911. Its secretary was Miss J. Halford, 4 Tavistock Square, London, WC. The Association aimed at bringing together all interests concerned with young children in order to promote the establishment of schools for mothers and advise them on their different circumstances. Affiliated societies paid an annual subscription ranging from five shillings to two guineas a year and individual members paid 2/6d.

Association of Inner Wheel Clubs in Great Britain and Ireland

The Association was formed in 1934 to promote friendship and to provide an opportunity for service in the community. The Inner Wheel Clubs, consisting of the wives of Rotarians, provide care for the sick and needy and the young, and raise funds for charitable purposes. Members have to be a wife or a widow of a Rotarian. The magazine *Inner Wheel* is issued three times a year.

Association of Lady Visitors

Founded by Adelaide, Duchess of Bedford in 1901, the Association was a later successor to Elizabeth Fry's Association of Female Prisoners in

Newgate. In 1922, members were allowed for the first time to visit boys as well as women.

(Prochaska, 1980)

Association of London Married Women Teachers

Established in 1909, this Association did sound work under the School Board for London and its successor, the London County Council, both of whom, unlike many other local education authorities, employed married women and made provision for maternity leave.

(Lowndes, 1950)

Association of Part-time Social Workers

The need for this organisation arose from the plight of married women social workers after the Second World War. The Association was formed in 1959 with the purpose of seeking to provide them with greater opportunities for part-time training and employment. The highest proportion of this category of workers were to be found in the Citizens' Advice Bureau and Family Welfare Association. Some valuable surveys on part-time employment were carried out by the Association in the early 1960s. It merged with the Association of Social Workers and the Social Work Advisory Service in June 1967.

(Wilmott, 1961)

Association of Post Office Women Clerks

This Association was formed in 1897 when women clerks' starting pay and annual increments were suddenly reduced. There were over 1,300 members by 1906. In 1913, the Association joined the **Federation of Women Clerks** and three years later, together with the **Civil Service Typists' Association**, formed the **Federation of Women Civil Servants**. When, in 1920, women were assimilated into the general grades of employment, it became part of the Civil Service Clerical Association.

Association of Registered Medical Women

Established in 1860, the Association held meetings on the first Tuesday in the month at the Medical Society of London, 11 Chandos Street, Cavendish Square, London.

Association of Teachers of Domestic Subjects

This Association, initially named the Association of Teachers of Domestic Science, was formed by a group of heads of domestic service training schools, superintendents and organising lecturers to represent their professional interests. Starting life as a Technical Sub-committee of the

National Union of Women Workers (later the **National Council of Women**) in 1896, the Association became an autonomous body the following year. It was open to all teachers holding diplomas or certificates in domestic subjects and the annual subscription was one shilling. Its aim was 'to form an educational centre for its members by the organisation of lectures to be given by experts at the Annual Conference, as well as, if possible, at other times'. By 1899 a new Executive of 30 members was elected to carry out the wishes of the Association. In 1909 direct membership was abolished and the Association became a federation of local associations. At the same time, the Association's title was changed from 'Domestic Science' to 'Domestic Subjects'. Students in their second year of training were also now eligible to join. The ATDS, in promoting the teaching of domestic subjects, acted as an agency for communicating resolutions and suggestions to the Local Education Authorities and the Board of Education on such topics as examinations in secondary schools, the teaching of science and pedagogical matters. In 1921, the first issue of the Association's house journal appeared. Because of shortage of funds, it then jointly published with the Institute of Handicraft Teachers a publication, *Practical Education and School Crafts*, which appeared six times a year. By 1928 the Association had its own journal, *Housecraft*. In 1932 there were 28 local associations. In 1983, it changed its name to the National Association of Teachers of Home Economics.

(Sillitoe, 1933)

Association of University Women Teachers

Established in 1883 by Anne Jemima Clough, the first principal of Newnham College, Cambridge and a vigorous campaigner for higher education for women. She served as its president until her death nine years later. The Association, which aimed at securing better conditions and terms of service for its members and promoting education generally, was 2,700 strong by 1914. When the Association of University Teachers was formed in 1919, it represented both men and women lecturers, and the Association was dissolved.

Association of Women Assistant Supervisors and Principal Clerks

The Association was founded in 1919 with a membership of 161. Its title was changed to the Association of Higher Grade Women of the Post Office Department in 1924 and it became affiliated to the Society of Civil Servants. After its merger with the Association for Higher Women Officers in the Civil Service that year, the membership rose to 3,000. It amalgamated with the Society of Civil Servants in 1940.

Association of Women Clerks and Secretaries

Established in 1903 under the title of the Association of Shorthand Writers and Typists, it initially included men; Sidney Webb was its first president. In 1912 membership, which consisted mainly of temporary employees, became for women only and its named was changed to the Association of Women Clerks and Secretaries. Many prominent feminists served as its officers, including Gertrude Tuckwell, and there were links with the **Women's Trade Union League.** Membership was restricted to highly qualified clerks and secretaries and the Association was less concerned with the broad mass of underpaid workers. In 1903 there were 64 members; by 1911, numbers had only risen to 104. By 1932, the Civil Service section of the Association had joined with the **Federation of Women Civil Servants,** which amalgamated with the Clerical and Administrative Workers' Union in 1941.

Association of Women Pharmacists *see* National Association of Women Pharmacists

Association of Women Pioneer Lecturers

The Association was formed for the purpose of providing lectures in centres not then covered by university extension or similar teaching. Competent women lecturers were employed who satisfied the requirements of a Lecture Sub-committee. It aimed at offering higher education to men and women engaged in regular work. Audiences included ladies in suburban districts, rural villages, night classes, institutes and polytechnics. Courses of ten lectures were normally offered and included history, literature, science, art and economics. In 1893, the secretary was Miss Edith Bradley at 13 Gray's Inn Square, London.

Association of Women Solicitors

The Sex Discrimination (Removal) Act of 1919 abolished disqualification on grounds of sex from admission to any incorporated society. The first woman solicitor, Carrie Morrison, was admitted to the Law Society in December 1922. The Association of Women Solicitors was formed in 1922 to protect the interests of women solicitors. It is active at the time of writing.

(Franz, 1965)

Association of Women Tax Clerks

When, in 1920, women were recognised as members of the established staff of His Majesty's Inspector of Taxes office, the Association of Women Tax Clerks was formed. It was merged in 1923 with the Association of Officers of Taxes, later the Inland Revenue Staff Federation.

Austral Club

Founded in 1905 at 44 St James's Court, Westminster, it provided a meeting place for women and men from Australia and New Zealand who were either resident in or visiting the United Kingdom. The Club also had an information bureau.

Auxiliary Territorial Service

With the approach of the Second World War and two days before Neville Chamberlain flew to Munich, the founding of the Auxiliary Territorial Service by Royal Warrant was announced, with power to recruit up to 20,000 women. Under its commandant, Dame Helen Gwynne-Vaughan, the Service was initially organised on a county basis and until the outbreak of war, some Companies served with the Royal Air Force. The Service provided a range of occupations, from forewomen and clerks to drivers, vehicle maintenance, radar operators, searchlight and anti-aircraft gun batteries personnel. In April 1941, the ATS were given equal military status with men, with its officers holding the King's Commission. The Princess Royal joined the ATS in August that year and was gazetted as Controller Commandant. With the country's increasing manpower problems, by 1942, for the first time in history, conscription for women was introduced and by the end of 1943 there were 212,000 members of the Service. Those aged between 19 and 40 were eligible for overseas service on a voluntary basis, though from January 1945, single women could be posted. The ATS served in France with the British Expeditionary Force, in North Africa and in other theatres of war, and many were awarded military honours. On 31 January 1949 the ATS became the **Women's Royal Army Corps**.

(Cowper, 1949)

Baptist Women's League

The League was established in 1908 with its first president, Mrs Herbert Marnham. It had three main objectives: to promote Home Mission work of the Baptist Union through raising funds on behalf of the Church and participate in evangelisation and colportage, to welcome Baptist women from other districts and assist girls in obtaining work and help pastors of rural churches. Candidates for admission were nominated by two members of the League. The subscription was one shilling per annum.

Bath Club

Founded in 1894 at 16 Berkeley Street, London, it afforded its members, both women and men, the opportunity for exercise throughout the year under cover, especially swimming. There were also squash courts and a

Morris tube rifle-range on the roof. The subscription included Turkish and other baths. In 1906 the club, which moved to 34 Dover Street, London, W, was non-resident and had 2,450 members.

Bee Club

This club was founded by Mrs Headlam at 19 Oakley Street, Chelsea, London, SW, in May 1901. The management was proprietary, but the club worked to be self-supporting only, not for profit. The entrance fee was five shillings and subscriptions varied between £1.8s and one guinea weekly. In 1904 the secretary was Miss Lilias Eaton of the **Beechwood Club**.

Beechwood Club

Founded in 1895 by Mrs Adair, it was a purely residential Club which was self-supporting and not for profit. It was situated at 6 Oakley Street, Chelsea. Subscriptions ranged from 18 shillings to two guineas a week and the entrance fee was five shillings.

Belfast Female Mission

Established in 1859, the Mission gave Bible classes for mill girls, made visitations to homes and held prayer meetings. In 1862 there were over 8,000 visitations. It was through the visitations that a scheme developed to provide nurses for the sick and it sent, where appropriate, cases to reform institutions. The Mission later developed sewing and cookery classes.

Berners Club

Originally the Working Women's Club, founded in 1869 at 9 Berners Street, London, W, it was one of the oldest women's clubs in the capital and perhaps the first to establish a meeting place for those seeking a career. Membership was extended to women from abroad. It had a well-equipped reading room with newspapers, periodicals and books, a conversation room and a visitors' room where members could receive friends. There was also a dining room where members could eat at moderate prices. It promoted frequent lectures and discussions. Subscriptions were raised from five shillings to ten shillings in 1871 and the club depended for its funds on its better-off members. In 1872 its name was changed to the Women's Club, then to the Berners Club. From 1872 to 1874 it appears to have shared its premises with the Central Committee of the **National Society for Women's Suffrage.** Its secretaries included Emma Paterson and Helen Blackburn. In the 1890s it moved to 22 Berners Street, the headquarters of the **Society for Promoting the Employment of Women**. The Club appears to have closed in 1898.

Birmingham Ladies' Association for Useful Work

Formed in 1874 by Miss Kenrick, then president of the Birmingham Ladies' Educational Association, and a large committee of ladies who wished to assist in promoting the education of women and girls, it originally had the management of the Cambridge Higher Local Examinations for Women and the Cambridge and Oxford Local Examinations for Girls. After the founding of Mason College (later, University of Birmingham) in 1880, it turned its attention away from the field of higher education to more modest activities. The Association delivered courses of lectures on hygiene and the laws of health to working women but also for teachers, pupil-teachers and servants. Research for the lectures were based on first-hand knowledge. For instance, out-patients' rooms in children's hospitals were visited to gather information for lectures on diseases in young children. The Association also opened a recreation room for business girls and organised country holidays for children.

Birmingham National Society for Women's Suffrage

Formed in 1868, it affiliated to the **National Society for Women's Suffrage**.

(Crawford, 1999)

Birth Control Investigation Committee

This was formed in February 1927 to research contraceptive methods and systematically collect data from existing clinics. It merged with the **National Birth Control Association** in 1931.

(Leathard, 1980)

Blackfriars Settlement *see* Women's University Settlement

Bristol National Society for Women's Suffrage

Formed in 1868, it affiliated to the **National Society for Women's Suffrage**.

British Association of Women Entrepreneurs

The main aim of this organisation, established in 1946, is to bring together women who are 'heads of business', that is, those who own or control an industry, company or business, to explore and advise on the means by which the rights and duties of women in business and domestic spheres can be reconciled and improved. Workshops and conferences are held and the organisation is part of the World Association of Women Entrepreneurs which hosts meetings in different countries. Its house journal is the *BAWE*

National News. The Association consists of over 200 members and associate members with associate membership available for those holding senior business positions.

British College of Nurses

Established in May 1926 by Ethel Bedford Fenwick and endowed by an anonymous donor, the British College of Nurses was seen as a rival organisation to the College of Nursing, later the **Royal College of Nursing**. Whereas the latter consisted of lay and national figures on its Council, the British College of Nurses was controlled entirely by nurses.

British Commonwealth League

It was formed in 1925 by the amalgamation of the British Dominions Women Citizens' Union (formerly the **British Dominions Women's Suffrage Union**) and the British Overseas Committee of the International Alliance for Woman Suffrage. Its original aim was to 'secure equality of liberties, status and opportunities between men and women in the British Commonwealth of Nations'. It affirmed a non-party, non-sectarian stance. The League acted both as a means of liaison for feminists in Commonwealth countries and as a means of raising issues of importance to women. Its concerns were various, and included such matters as the nationality of married women, mixed marriages, comparative rates of pay, prostitution and sexual oppression, and the rights of aboriginal Australian women, issues that were debated vigorously at its annual conference. Prominent members included Margery Corbett Ashby, Chave Collisson, Myra Sadd Brown and Alice Hemming. In 1963 the League recognised developments in the evolution of the Commonwealth and changed its name to the Commonwealth Countries League, but it has continued vigorously to pursue the same broad objectives, at least up to the time of writing.

British, Continental and General Federation for the Abolition of Government Regulation of Prostitution: British Branch

At a conference of those working for the repeal of the Contagious Diseases Acts at York on 25 June 1874, Dr Collingwood, together with Josephine Butler and other leading campaigners against the Acts, decided that contact should be made with those on the continent of Europe opposed to the system of state regulation of prostitution. In consequence, the British, Continental and General Federation was formed in Liverpool on 19 March 1875, with James Stansfeld as president, Josephine Butler as secretary and Henry Wilson as corresponding secretary. Subsequently William Crosfield was appointed treasurer, and James Stuart took over from Henry Wilson. The British branch was based in Liverpool, at 248 Park Road, with a London office managed by Frederick Banks at 27 Great George Street, Westminster. After the repeal of the Contagious Diseases Acts in 1886 the Branch concentrated mainly on opposing regulation of prostitution in the British Empire, particularly India, and on the re-enactment of the Contagious Diseases Act (India) in 1897 it revived the repeal periodical the *Shield*. In 1890 it had already changed its official name to the British Committee for the Abolition of the State Regulation of Vice in India and throughout the British Dominions. In 1915 it merged with the **Ladies' National Association for the Repeal of the Contagious Diseases Acts** to form the **Association for Moral and Social Hygiene**. The British, Continental and General Federation at this time changed its name to the International Abolitionist Federation.

British Dominions Women's Suffrage Union

Originally formed in New Zealand in 1913 by Margaret Hodge and Harriet Newcombe, it was established in London in July 1914. It attracted the active support of such prominent and different suffragists as Millicent Fawcett, Sylvia Pankhurst and Emmeline Pethick-Lawrence. It held conferences in London in 1916 and 1918. In 1918 it was re-named the British Dominions Women Citizens' Union; its secretary was then Daisy Solomon. In 1922 it amalgamated with the British Overseas Committee of the International Woman Suffrage Alliance, which in turn in 1925 joined the **British Commonwealth League**.

(Crawford, 1999)

British Federation of Business and Professional Women

Formed in 1935 in conjunction with the International Federation of Business and Professional Women's Clubs (in existence since 1930) on the initiative of the **Women's Advertising Club of London** and the **Women's Engineering Society**, its first chairman was Caroline Haslett, and its first president Lady Astor. Its objectives were to promote and safeguard the interests of business and professional women, to establish a closer co-operation between them, and to secure effective and consistent co-ordination of action upon matters affecting their mutual interest, as well as to promote and maintain professional standards. The Federation was especially prominent in the Second World War, when it campaigned vigorously over such issues as woman power, equal compensation for war injuries, equal pay, etc. Its organ from 1940 was *Women at Work*, which continued until 1951. The Federation was finally dissolved in 1969. It is not to be confused with the **National Federation**

of Business and Professional Women's Clubs of Great Britain and Northern Ireland.

(Faulkner, 1969)

British Federation of University Women *see* **British Federation of Women Graduates**

British Federation of Women Graduates

At a meeting of 17 women at Manchester High School in March 1907, a proposal was made to form a Federation of University Women. One of its leading members was Ida Smedley, then an assistant lecturer in chemistry at Victoria College, Manchester and holder of a London DSc degree. It was agreed that the Federation should 'afford a means of communication and of united action in matters affecting the interests of women'. The Federation would encourage women's representation on public bodies, work for the removal of sex discrimination in employment and express a united opinion by university women on subjects of special concern to them. From 1908 the Federation became based in London and in its first two years took up many causes. In Manchester, the Federation supported one of its members as a candidate for the Town Council, tried to secure the appointment of a woman as Resident Medical Officer at the Infirmary and expressed concern at the small number of women on the city's Education Committee.

The Federation was split on the question of suffrage and the issue played no major part in its deliberations. Drawing rooms and public meetings were used for promoting campaigns, such as poor law administration and temperance reform, and contact was made with similar bodies such as the **Association of University Women Teachers** and the Irish Graduates' Association. Another of the Federation's concerns was to promote and safeguard the interests of women in academic work at a time when the majority of university posts were filled by men. Money was collected by members for the annual award of a Prize Fellowship to assist promising students. The Federation widened its scope in 1933 when it decided to help university graduates who wanted to enter professions and vocations 'beneficial to the public interest'. A Federation of Organisations for the Training and Employment of Educated Women was subsequently set up. In 1925, the administrative grade in the Civil Service was opened to women as well as men and two women were successful that year. A more difficult case was the Foreign and Diplomatic services, which were not open to women until 1946; even then a marriage bar was enforced. Entry into industry and commerce was also encouraged. A majority of the Federation's members were engaged in the field of teaching and it fought a long battle for equal pay. In 1955 the government finally announced that equal pay in the Civil Service would be introduced over a seven-year period, and this precedent was followed by the teaching profession. In 1926, the Federation had acquired Crosby Hall, Chelsea, part of which incorporated a fifteenth-century hall, which was used as an international hostel for scholars. It is now known as the British Federation of Women Graduates; its address at the time of writing is 25 Great James Street, London,WC1.

(British Federation of University Women, 1957)

British Housewives' League

This was formed by Irene Lovelock and Alexandra Landau in July 1945 in London as a protest against austerity measures, particularly food rationing, which were being extended by the Labour government. Similar initiatives outside London, particularly on Merseyside, eventually joined with the League. It organised petitions and demonstrations against queuing and bread rationing. It was widely perceived as being associated with the Conservative Party, and indeed some local branches appear to have acted as front organisations for the Conservatives, but there were no formal links at national level; in fact the Conservative Party seems to have considered its political stance too extreme. Its original chairman, Mrs Bloice, was displaced by Dorothy Crisp in March 1947. She attempted to turn the League into an independent electoral force, but Mrs Bloice, encouraged by Lord Woolton, the Conservative Party chairman, attempted to start a rival organisation called the Women's United Front, which was supposed to be non-party, but, in the words of Conservative Central Office, 'entirely friendly to us'. By December 1948 both the League and the Front seem to have come to the effective end of their activity, although the League is still in existence at the time of writing.

(Hinton, 1994)

British Industrial League and United Gentlewomen's Handicraft and Home Industries Exhibition

The object of the Guild was to support the home-industry movement and create an interest in art handicrafts, by which a large number of ladies endeavoured to support themselves in the early part of the twentieth century. Exhibitions were held which promoted the work of its members. Among its patronesses shortly before the First World War were the Duchesses of Norfolk, Devonshire and Rutland.

British [Ladies'] Female Emigrant Society

Not in fact a Society to assist women's emigration, this body was founded in about 1850 to protect those women whom they considered ill-advised enough to emigrate. It was also known as the British Matron Society. Its secretary was Miss Caroline Tipple, on whose death in 1878 its work was taken over by the **Colonial Emigration Society**.

(Hammerton, 1979; Monk, 1963)

British Matron Society *see* **British [Ladies'] Female Emigrant Society**

British Society of Ladies for Promoting the Reformation of Female Prisoners

In 1813 the Quaker, Elizabeth Fry, visited Newgate Prison in London and was disturbed at the conditions endured by the female prisoners. Eight years later, together with 12 other Quaker women, she founded the Association for the Improvement of Female Prisoners in Newgate. Besides improving their physical conditions, the inmates received systematic religious instruction and elements of education by ladies of the Association. Its success led to the establishment of a national body, the British Society of Ladies for Promoting the Reformation of Female Prisoners. Fry advocated the provision of separate prisons for women and the institution of further Associations in Britain.

(Luddy,1995)

British Women Pilots Association

Founded in 1955, with its headquarters at Rochester Airport, Kent, for women pilots interested in gaining a better understanding of aviation. It also offered career counselling services.

British Women's Emigration Association

This was the name taken in 1901 by the former **United British Women's Emigration Association**. Its organ was the *Imperial Colonist*, which ran 1902–27. Like its parent body, its main concern was placing domestic servants and protecting them in their travels. Two members of its council opened a Colonial Training Home, first at Leaton, then at Stoke Prior, which 'devoted itself to training women of superior standing for the duties of Companion Help'. After the South African War, a sub-committee, the South African Expansion Committee, became a virtually autonomous organisation, under the title of the South African Colonisation Society. By 1905, the Society had assisted 8,000 people to settle in different parts of the Empire. Together with its parent body, it merged in 1919 with the **Colonial Intelligence League for Educated Women** to form the **Society for the Oversea Settlement of British Women**.

(Hammerton, 1979; Monk, 1963)

British Women's Temperance Association

It was founded in 1876 under the presidency of Mrs Edward Parker to organise women to encourage temperance by education and other means, and to agitate for the restriction of sales of alcohol. In addition it provided activities for children from the age of seven and young people up to the age of 30 which highlighted the danger of alcohol. Weekend and summer schools were arranged as well as examinations and competitions. It was affiliated to the World Women's Christian Temperance Union. Its organ was originally the *British Women's Temperance Journal*; in 1892 this changed to *Wings*, but at this time there were divided counsels within the Association. Lady Henry Somerset had become prominent in the Association, and had been attempting to draw it into an alliance with the women's suffrage movement along the lines already operating in the USA. This, together with Lady Henry's style of leadership, did not meet with the approval of many members, and in 1893 a substantial number of them broke away to form the Women's Total Abstinence Union, taking with them the weekly *Wings*, Lady Henry having already taken over editorship of the feminist weekly the *Woman's Herald*, which now became the organ of the Association. In 1894 it became the *Woman's Signal*, still officially the Association's organ, but in fact under the editorship and proprietorship of Florence Fenwick-Miller. In 1896 the Association started its own paper, the *White Ribbon*. In 1925 the Association and the Women's Total Abstinence Union resolved their differences and amalgamated under the name of the **National British Women's Total Abstinence Union**, which is the name it bears at the time of writing. The Union's concerns include the spread of gambling and the need for moral welfare; it is also an active supporter of world peace.

(VanArsdel, 1979)

Build the Body Beautiful League *see* Women's League of Health and Beauty

Camelot Club

This club was established at 29 Queen's Square, London, WC in 1898 by the Hon. Coralie Glyn and was intended as a convenient meeting place for working professional women. There was no entrance fee and the subscription was five shillings. Election was by committee. Excursions and social gatherings were arranged and the club remained open on Sundays from 11 a.m. to 9 p.m. In 1901, there were 70 members.

Catholic Women's League

The League was founded in 1906 with its headquarters at 48 Great Peter

Street, London, SW1, as a national, non-political organisation for charitable, religious and intellectual work. It co-operated with both Catholic and non-Catholic societies for providing lectures, study facilities and libraries as well as baby clinics, **Girl Guides**, residential homes and clubs for working girls and emigration, and gave help to refugees and victims of war persecution. Issues affecting women were raised on public bodies by members of the League at national and international levels. It published a journal, the *Catholic Women's League Magazine*.

Catholic Women's Suffrage Society

Formed in 1911 to campaign for women's suffrage and feminist issues in general among Catholics, it eventually became subject to the discipline of the Catholic hierarchy over its name, and in 1923 it became the **Saint Joan's Social and Political Alliance**. Its organ was the *Catholic Suffragist* (1915–18), which then continued as the *Catholic Citizen*.

(Mason, 1986; Parnell, 1961)

Central and East of England Society for Women's Suffrage

In the 1888 split within the **National Society for Women's Suffrage**, this was the London-based branch supporting the Central Committee. It was one of the constituent societies which formed the **National Union of Women's Suffrage Societies** in 1897, and in 1900 it was merged in the **Central Society for Women's Suffrage**.

(Crawford, 1999; Rover, 1967)

Central and Western Society for Women's Suffrage

In the 1888 split within the **National Society for Women's Suffrage**, this was the London-based branch, at 39 Victoria Street, London, SW1, supporting the Central National Society for Women's Suffrage. It was one of the constituent societies that formed the **National Union of Women's Suffrage Societies** in 1897, and in 1900 it was merged in the **Central Society for Women's Suffrage**.

(Rover, 1967)

Central Committee of the National Society for Women's Suffrage
see **National Society for Women's Suffrage**

Central Committee on Women's Employment

This Committee was set up at the beginning of the First World War 'to consider and from time to time report upon schemes for the provision of work for women and girls unemployed on account of the war'. The Committee had begun life as the Queen Mary's Needlework Guild and then changed its name to the Queen's Work for Women Fund. The original scheme attracted many unpaid volunteer ladies which aggravated the unemployment situation of dressmakers, and as a result of protest by women trade unionists, they were given places on the Central Committee. Mary Macarthur was honorary general secretary with Susan Lawrence, Margaret Bondfield and Dr Marion Philips acting as committee members alongside society ladies. The Committee's main tasks were to find employment for working women in industry and help local representative committees to provide relief work for the unemployed. Many clothing firms provided employment though the centrally-fixed wage rate of 3d. an hour for a 40-hour week was criticised. Training courses were also provided in London in dressmaking, tailoring and millinery as well as in clerical skills. By 1915 the problem was largely solved by the need for women to take over many jobs formerly filled by men in factories and in the burgeoning war industries.

Central Employment Bureau for Women

Established in 1899, the Bureau provided employment advice for women mainly through its organ *Women's Employment*. Its publishing arm was the Women's Employment Publishing Company. It appears to have ceased activity about 1975.

Central National Society for Women's Suffrage *see* National Society for Women's Suffrage

Central Society for Women's Suffrage

Despite the formation of the **National Union of Women's Suffrage Societies** in 1897, the two formerly rival Central Societies had continued in existence. In 1900 they were merged in the new single Central Society, based in London, which in 1907 became the **London Society for Women's Suffrage**.

(Rover, 1967)

Central Women's Advisory Committee

A group of Conservative women MPs, conscious of their lack of involvement in the formulation of policy and their minor role in parliamentary matters, established the Central Women's Advisory Committee in the early 1920s. It was not until 1928 that it was officially recognised. After the Party's electoral defeat in 1945, its title was changed to the Women's National Advisory Committee (WNAC) and the separate men and women's constituency branches were abolished. Three sub-committees, consisting of women MPs , life peeresses and women officers of the National Union of the WNAC were formed, Parliamentary, General Purposes and Outside Organisations. The Parliamentary sub-committee considered proposed legislation which affected the interests of women, such as the issue of equal pay, wives' rights in divorce, joint signatures of husband and wife on income-tax returns and unfair exclusion of women from jury service.

Chartered Society of Physiotherapy

The Society was formed in 1894 after the *British Medical Journal* published a series of articles which revealed the existence of large-scale massage scandals which were disguised as physiotherapy. Two young nurses, Miss (later Dame) Rosalind Paget and Miss Lucy Robinson, together with some medically trained friends, decided to form a Society of Trained Masseuses which granted certificates of competence for successful candidates together with a code of practice for genuine practitioners. The first meeting consisted only of members of this group, but they were subsequently joined by staff from hospitals and private schools. The Society set examinations and educational standards, inspected training schools and worked to improve the status of its members in the medical hierarchy. By the outbreak of the First World War, the Society consisted of 2,657 members. In 1900, its name was changed to the Incorporated Society of Trained Masseuses and in 1920 it amalgamated with the Chartered Society of Massage and Medical Gymnastics. It assumed its present title, the Chartered Society of Physiotherapy, in 1942. By 1944, it had over 26,000 members, consisting of both men and women.

(Ward, 1980)

Children's Jewel Fund

From 30 July 1914, the Local Government Board and the Board of Education made grants in aid for local maternal and child welfare work. The local authorities were urged to make provision for expectant and nursing mothers and young children. The spread of child welfare centres received a boost with the founding, in 1917, of the Children's Jewel Fund, by a group of public spirited ladies who wished to encourage the starting of new centres and extend the work of existing ones. An office was opened in New Bond Street, London at the beginning of 1918 and an appeal was launched for contributions, either in money, jewels or other valuables. 'A jewel for a baby's life' became one of its mottoes. There was an overwhelming public response; family heirlooms, wedding rings and kissing cups were among thc gifts received and Queen Mary and Queen Alexandra contributed generous sums of money. A performance of

Manfred at Drury Lane was mounted to raise funds and a sale of jewels in 1918 brought in the sum of £26,000. Thanks to a brilliantly run propaganda campaign by the committee of the Fund, by the time it closed in 1920 it had raised £70,000 for distribution to 164 institutions.

(McCleary, 1935)

Christian Alliance of Women and Girls

An Evangelical organisation, its aim was to win over young women to religion. An extensive programme for education and recreation was provided, including drill, and there were hostels either for residential or holiday purposes. During the Second World War, its Service Girls' Canteens and Recreations Centres for the Armed Forces were popular. In 1946, there were 9,000 in the movement.

Christian Women's Union Conference

This Conference, founded in 1879, was intended to embrace all Christians, not confining its efforts to those connected with churches. It carried out rescue work, held Bible classes and dinner-hour meetings in city warehouses for working girls, promoted temperance and, at its annual three-day conference, discussed such matters as missionary work overseas and working amongst prisoners. It specifically stated that its object was not to supersede men in the Church but rather to find suitable tasks for Christian women to undertake.

Church Extension Association

The Association, which began in 1864, had as its main objective the assisting of home missions and the establishment of ragged schools. Six years later, the Community of the Sisters of the Church was formed which devoted itself to the work of the Association. The Association opened girls' orphanages, organised on family lines, a boys' home and a foundling home, 15 large day schools and a teachers' training college. Other amenities included a seaside convalescent home, a night shelter for men and an accident hospital at Rotherhithe for dockers, with facilities for 2,600 outpatients.

Church Imperial Ladies' Club

Situated at 2 Army and Navy Mansions, Victoria Street, London, it was in existence in 1916. The entrance fee was two guineas, subscriptions were two guineas and for country members, one guinea.

Church League for Women's Suffrage

Founded in 1909 by the Rev. Claude and Mrs Hinscliffe, it campaigned for women's suffrage within the Church of England. In 1913 it claimed a total of 66 local branches, with an approximate membership of 3,000, including five bishops. Prominent members included Ursula Roberts, Helen Hanson and Maude Royden. From 1912 it ran a monthly paper, and by this time its banner depicted St Margaret of Antioch. In 1917 it became the **League of the Church Militant**. The League also had an Irish Branch, secretary Miss Stack, at Castlewood House, Rathmines.

(A. J. R., 1913; Crawford, 1999)

Church Missionary Society: Ladies' Union

Formed in the nineteenth century in connection with the Society for the Propagation of the Gospel, chiefly to raise large sums of money for carrying on missionary work abroad and providing female teachers for the instruction of native women and children in the missions of the Society. Its original title was the Ladies' Association for the Promotion of Female Education in India and other Heathen Countries. It also helped female mission schools by providing suitable clothing and maintenance for its boarders. In the early 1890s, there were 850 associations of women who raised £31,000 per annum for the Association. Meetings were periodically held to report on information received about the activities of the Society at home and abroad. Besides the London Union, there were several others formed in provincial towns. The Gleaners' Union for Prayer and Work and the Sowers' Band, aimed at interesting children in the Society, were organised by women. The Association sent over 100 female missionaries abroad from 1887, and over 200 wives of missionaries participated in the work. It also maintained 60 schools, orphanages, hospitals and other institutions which were staffed by these ladies. The Association changed its name to the Women's Missionary Society in 1894 and added medical work to its activities three years later. It merged with the Society for the Propagation of the Gospel in 1904 but a committee of women's work was subsequently formed.

Church of England Women's Help Society

Begun in Colchester in 1879, the Society's aims were to help both married and single women to lead fuller Christian lives. Members had to be regular communicants of not less than a year's standing and of good character. Membership was attained by progressive stages for those who adopted the appropriate Rules of Life. For this purpose, a number of activities were organised to draw members together. These included mothers' meetings, lending libraries, penny banks and sick clubs; lodging houses were provided for women employed in factories and warehouses. An annual meeting of working women and girls was held in St Paul's

Cathedral. By 1906 there were over 150 branches in existence including nine in the Colonies. Girls aged between 14 and 17 were known as Little Sisters and wore a badge with the word 'Persevere' on it. The Society's headquarters was at Church House, Westminster. The Society, like the **Girls' Friendly Society**, operated through the 27 dioceses in England and Scotland.

Church of England Zenana Missionary Society

Deriving from the Church Missionary Society, this was founded in 1880 with the original object of taking the Gospel to women-only environments (Zenanas) in India. Its scope soon broadened to more generally educational and medical, and its geographical area of concern was expanded to include Ceylon (Sri Lanka), China and Japan. Its organs were *Homes of the East*, *Daybreak* and the *Indian Female Evangelist*. It remained active until 1957.

Church of Scotland Guild

Established in 1887 the Guild was started as a movement within the Church of Scotland to encourage women to express their faith in worship, prayer and action. A national conference is held annually and activities are locally arranged. In 1997 there were 54,300 members.

(Women's National Commission, 1997)

Church of Scotland Women's Association for Home Missions

This organisation trained home mission workers in the Church of Scotland. The course lasted one or two years and consisted of Bible study, district visiting or nursing. Home Mission workers were either deaconesses of the Church of Scotland or Parish Sisters, many of whom worked in the slum areas of large cities.

Civil Service Typists' Association

This Association, formed in 1903 and established in 1911, represented 90 per cent of the 601 women Civil Service typists; from its beginning it pressed for equal conditions for men and women. In 1913 all the Civil Service women's clerical associations amalgamated and formed the **Federation of Women Clerks**; three years later, it joined with the Civil Service Typists' Association to form the **Federation of Women Civil Servants**. At the outbreak of the First World War, there were over 13,500 members. In 1920 it left the Federation and transferred to the Civil Service Clerical Association.

Civil Service Woman Suffrage Society

The aim of this Society, founded in 1911, was to bring together women

Civil Servants belonging to different suffrage societies in order to use their influence in their professional capacity. The honorary secretaries in 1913 were B. M. Rendle and E. A. Smyth, by which time it had a membership of about 600.

(A. J. R., 1913; Crawford, 1999)

College by Post

Founded by Miss Mary Petrie, John Stuart Mill Scholar in Philosophy, University College, London, in 1881, this organisation aimed at helping young women seeking to improve their education by means of correspondence. Between 200 and 300 lecturers and teachers from Girton and Newnham Colleges, Cambridge, Lady Margaret and Somerville Halls, Oxford, University College, London and Westfield College, Hampstead, Ladies' College, Cheltenham, as well as from other institutions, volunteered to take part in this work. By 1893 almost 4,000 students from the United Kingdom, the Continent and the Colonies had enrolled. Instruction was provided free on condition that the student devoted half an hour each day to Bible study.

(Burdett-Coutts, 1893)

Colonial Emigration Society

Formed in 1884 from the Marylebone Branch of the **Women's Emigration Society**, and attracting most of that Society's members. It carried on its work, and that of the **Female Middle-Class Emigration Society**, with a special emphasis on educated and qualified women, until 1892, when it was merged with the **United British Women's Emigration Association**. Its organiser was Mrs E. L. Blanchard.

(Hammerton, 1979; Monk, 1963)

Colonial Intelligence League for Educated Women

Started in 1910 by the Hon. Mrs Norman Grosvenor as the Committee of Colonial Intelligence for Educated Women, this body was specifically aimed at 'poor ladies', usually with professional qualifications beyond teaching. In 1919 it merged with the **British Women's Emigration Association** to form the **Society for the Oversea Settlement of British Women**.

(Hammerton, 1979; Monk, 1963)

Colonial Nursing Association

The object of the Association, which was set up in the late nineteenth century, was to provide trained hospital and private nurses for the Colonies and small British communities in foreign countries, to guarantee their salary and passage money and to financially assist colonies and communities abroad who were unable to bear the entire expenses of maintaining the nurses. In 1916, its patron was Princess Henry of Battenberg, and the honorary secretary, Lady Susan Hicks Beach.

Columbia Club

Founded in December 1892 at 19 Lexham Gardens, London, W, this was a residential Club with members elected by a committee. Subscriptions were a guinea, as was the entrance fee, the latter being remitted in the case of wives and unmarried daughters of clergymen and officers in the Army and Navy. In 1904 the secretary was Mrs Gordon Haynes.

Committee of Colonial Intelligence for Educated Women *see* **Colonial Intelligence League for Educated Women**

Commonwealth Countries League *see* **British Commonwealth League**

Communist Party of Great Britain: Women's Network

This organisation was formed in 1920. It encouraged the involvement of women in the Communist Party and advises on women's issues. The Network has been active in the fields of human rights and peace initiatives, the environment and child care. In 1993, there were 18,000 members.

Conservative and Unionist Women's Franchise Association

Founded in 1909 by Lady Knightley of Fawsley, and with an impressive array of eminently titled vice-presidents (in 1913, there were three duchesses, three marchionesses, four earls and many countesses and viscountesses), it organised Conservative suffragists and propagandised for women's suffrage among Conservative party members and voters, and supported Conservative and Unionist candidates who were in favour of the enfranchisement of women. In 1913 it had 61 local branches, and had its own monthly paper, which ran till 1916. It continued its activities until at least 1924. Its organs were the *Conservative and Unionist Women's Franchise Review* and *Monthly News of the Conservative and Unionist Women's Franchise Association*. The Association had an Irish Branch with its headquarters at 51c Dawson Street, Dublin. In 1913, its president was the Countess of Fingall and the secretary was Miss Patton.

(A. J. R., 1913; Crawford, 1999; Doughan and Sanchez, 1987; Gordon, 1999)

Conservative Women's National Committee

Established in 1904, the committee consisted of women members of the

Conservative Party in England and acts as adviser to the Party on matters of common concern to women. It works for the election of members of the Party at both national and local levels, and encourages women to stand for Parliament. Its headquarters are at 35 Smith Square, London, SW1.

Conservative Women's Reform Group

Towards the end of the Second World War, two Conservative MPs, Thelma Cazalet-Keir and Irene Ward, who aspired to advance women's issues within the Party, formed the Conservative Women's Reform Group. They were the authors of a pamphlet, *When Peace Comes*, which considered the role of women after the war. It was aimed mainly at professional women who were not politically committed and might be persuaded to join the Conservative Party.

Consultative Committee of Women's Organisations

It was established by Lady Astor in 1921 to co-ordinate efforts with such bodies as the **National Union of Societies for Equal Citizenship**, the **Women's Freedom League** and the **Saint Joan's Social and Political Alliance** to lobby for legislation to improve women's status (and incidentally to give credibility to Lady Astor's claim as the sole woman MP at Westminster at the time to speak for women in general). It also campaigned at parliamentary elections and by-elections, and was particularly involved in the campaign for equal franchise. However, in 1928 it dissolved over the question of protective versus restrictive legislation.

(Smith, 1998)

Council for the Amelioration of the Legal Position of the Jewess

This Council, which flourished in the 1930s, was founded by Miss L. Hands, 57 Portdown Road, London, W9. It had four objectives: to press for measures of relief which would help women who under then existing laws would be liable to suffer serious hardship; to counsel with Jewish communal workers and organisations throughout the world concerning the legal disabilities of Jewish women; to consult with Jewish scholars as to the best methods to employ in order to obtain the desired results; and ultimately to lay these considered demands before a universal assembly of Orthodox rabbis.

Council of Married Women

It was formed at a meeting inspired by Helena Normanton QC, and held at the Goring Hotel, London, SW1, on 31 March 1952. The first officers of the Council were Helena Normanton (president), Doreen Gorsky (chairman), Lady Helen Nutting (deputy chairman) and Evelyn Hamilton (honorary secretary and treasurer). Its objects were to support the stabilisation and dignity of marriage as an institution; to promote the further equalisation of the sexes in marriage and its consequences; and to initiate legislation for legal and economic justice for wives, mothers and widows, for women who have obtained decrees of dissolution of marriage, separation orders, and for children and orphans. It was distinguished from other societies working for equal citizenship for women by concentrating on obtaining equal economic status for women in marriage.

Its separation from the **Married Women's Association** was occasioned principally by the opposition within the Association to the evidence prepared by Helena Normanton for presentation to the Royal Commission on Marriage and Divorce, as well as what was considered to be the irresponsible behaviour of some members of the Association, relating to the Korean War. By 4 June 1952, the Council had 31 full members, three associate members and one Group of 15 members. On 16 November 1953, Doreen Gorsky resigned as chairman because of the demands of her new appointment at the BBC, and Lady Helen Nutting was elected chairman, a post she held until the Council's dissolution. Shortly thereafter Helena Normanton resigned as president because of a personality conflict with the chairman. After the findings of the Royal Commission appeared in 1956, the Council concentrated on attempts to obtain a Private Bill through Parliament embodying their main plank of the economics of marriage, namely the statutory right of a wife to a portion of the family income. Meetings became extremely infrequent during the 1960s, the main business of the Council apparently being carried on by the chairman. The Council was wound up on 31 May 1969.

Council of Women Civil Servants

This Council was formed in 1920 and was open for membership to all women who were in the Administrative, Professional and Executive Classes of the Civil Service. Its objective was to ensure that equality of opportunity was achieved in all departments. The Council presented evidence to a number of Royal Commissions concerned with pay and conditions of service of women workers and it made contacts with many other outside women's organisations. The Council was dissolved in December 1968.

Cruse Clubs Ltd

This counselling service for widows and their children was incorporated in 1959. Its aims and objectives included collecting and publishing information about social services, statutory or otherwise, useful to widows; assisting the formation of Cruse Clubs to overcome widows' isolation; to inform public opinion about the needs of widows; to undertake research on the problems of widows; and to assist in providing holidays for widows who could not otherwise afford them. It took its name from the widow's cruse, the traditional name for the vessel filled with oil for a poor widow by the prophet Elisha. The original chairman and honorary director was Mrs Margaret Torrie; patrons included Dr Leslie Weatherhead, Lady Isobel Barnett, Marghanita Laski, Rupert Davies and Katharine Whitehorn. By the 1980s it was widening its scope to include widowers and the bereaved generally. Its organ was the *Cruse Chronicle.*

Cymric Suffrage Union *see* Forward Cymric Suffrage Union for Men and Women

Denison Club

Founded in 1886 at 15 Buckingham Street, Adelphi, Strand, London, WC, for ladies and gentlemen interested in charitable and social subjects. Meetings and conferences on these areas were held there to which outsiders were admitted. The Club was non-residential. The minimum subscription was 10 shillings. By 1901 it had 130 members. At about this time it moved to Denison House in Vauxhall Bridge Road, London, SW. It was still active in 1909.

Dorothy Restaurant

A restaurant for the exclusive use of women, opened by the Ladies' Restaurant Association at 81 Mortimer Street, London, W, and which was flourishing in 1880. It provided dinner, which consisted of two joints, bread and two vegetables, at a cost of 8d., served between midday and 3 p.m. An à la carte menu was also available between the hours of 8 a.m. and 8 p.m. except between midday and 3 p.m. Separate tables were provided and it was advertised as for 'young ladies engaged in business during the day'.

Dublin Ladies' Clothing Society

Founded in 1849 under the patronage of Archbishop Daniel Murray, the Society provided cheap or free clothing for the poor. Women members' activities were limited by male members of the Society of St Vincent de Paul, who handed out tickets to selected applicants. Despite its name, the women's role was confined merely to distributing clothing and raising funds for the Society.

Dublin Ladies' Sanitary Association

Developed from the National Association for the Promotion of Social

Science, this Association was formed in 1861 to 'improve the physical health of the poor, and to dispel the ignorance of the laws of health that exists among all classes'. Membership was open to both Protestant and Catholic women and they were instructed in aspects of health which they could then use on their home visits. It ceased to be active by 1890.

(Luddy, 1995)

Dublin Visiting Mission

An evangelical organisation promoted by the Irish Church Mission in Dublin in 1849, it employed scripture readers to visit the homes of the poor, ran Sunday Schools, prayer meetings and Gospel meetings. One of its leading members was Mrs Ellen Smyly, who also opened a home for destitute children.

East London Federation of Suffragettes *see* Women's Social and Political Union

East London Jewish Girls' Club

Founded in 1886 by Lady Magnus, pioneer of the girls' club movement and wife of Sir Philip Magnus.

East London Ladies' Club

The Club, located at Toynbee Hall, was started in May 1899 as a meeting place for ladies of every denomination engaged in philanthropic and social work in East London. It included a rest room and other facilities and was open every day except Sundays. The subscription was five shillings per annum. In 1906 it had 22 members.

East London Nursing Association

A similar but smaller scale version of Mrs Ellen Ranyard's **London Biblewomen and Nurses' Mission**, this Association was set up in 1868 and was the earliest example of district nursing for the poor in London. A trained nurse was located in each parish, with lodgings supplied from local funds. Superintendence was carried out by matrons living in the immediate neighbourhood. By 1903 there were 29 nurses working under four matrons.

Edinburgh Association for the University Education of Women

Originally the **Edinburgh Ladies' Educational Association**, its object was to provide a substitute for the university education of women modelled mainly on the Arts Faculty of Edinburgh University. The University participated in its work in so far as it granted certificates in Arts subjects to students who had attended a certain number of classes of the Association and had passed the necessary examinations. The lecture hall was at 15 Shandwick Place, Edinburgh. In 1888, its honorary secretary was Mrs Louisa Stevenson.

Edinburgh Ladies' Educational Association

This Association was founded in 1867 with the objective of advancing the higher education of women. As a means towards this end, the Association attempted to establish a lecture scheme. The initial series of 30 lectures on the history of English literature was given by Professor David Masson. The president was Lady Dunfermline; the founders were Mrs (later Dame Sarah) Mair, Mrs Lindsay, Mrs De Lacy Evans, Mrs Daniell, Mrs Ranken and Mrs Crudelius. The Association appears to have ceased activity some time after 1869.

Edinburgh Ladies' Emancipation Society

Founded in 1833, and devoted to promoting the universal extinction of slavery, the Society was still active in 1861, when its membership stood at just under 500. Its president was Mrs Wigham, its treasurer Mrs Duncan and its joint secretaries, Mrs Lillie and Miss Wigham. It held monthly meetings in the Religious Institution Rooms, 6 York Place, Edinburgh. There was a small library of anti-slavery materials for the use of members. The subscription was two shillings and sixpence. Its motto was 'Am I not a man and a brother? Am I not a woman and a sister?'

Edinburgh National Society for Equal Citizenship *see* Edinburgh National Society for Women's Suffrage

Edinburgh National Society for Women's Suffrage

Formed in 1867, it joined with the London and Manchester National Societies to form the **National Society for Women's Suffrage** later that year, while continuing an independent existence. Prominent members included Agnes McLaren and Jane Taylour. In 1872 it associated with the central committee of the National Society for Women's Suffrage, and in the 1888 split, with the Central National Society. In 1897 it affiliated to the **National Union of Women's Suffrage Societies**, and in 1918 it became the Edinburgh National Society for Equal Citizenship.

(Crawford, 1999; Leneman, 1995)

Edinburgh University Women's Suffrage Society

This was probably the only university suffrage society to produce its own paper, *The Only Way*. Single issues from 1909 are held in the Fawcett Library and the Bodleian Library.

Electrical Association for Women

This Association was established in 1924 under the directorship of Caroline Haslett, a former suffragette and organiser of the **Women's Engineering Society**. It had as its objective the safe and efficient use of electricity in the home and an improvement in electrical appliances. It held short courses for the general public, electrical cookery classes, certificated courses for schools and colleges, mounted travelling exhibitions and awarded certificates for teachers and demonstrators who passed the appropriate examinations. The Association became a successful political pressure group in government circles as well as within the electrical and building industries. It was able to secure appointments in the electrical industry for a few women in the 1920s and campaigned for cheaper electricity so that poor people were able to afford it. Branch associations were formed outside London, often chaired by a titled lady. By 1939 there were 80 associations. The illustrated house magazine was called *Electrical Age for Women*.

(Davidson, 1982)

Emerson Club

This Club was founded in 1911 at 19 Buckingham Street, Strand, London. By the 1920s, it was the home of the Women's Group of the Ethical Movement, at which time prominent shareholders included Myra Sadd Brown, Alison Neilans and Dr Elizabeth Knight. When it closed in 1928 its address was 1 Little St George Street.

Empress Club

Established in 1897 at 32 and 35 Dover Street, Piccadilly, to mark the Queen's Jubilee. One of the most luxuriously appointed clubs in London (it included a smoking gallery and a smoking room, a tickertape for stock-market news, a room in which servants could be interviewed and 'one of the best orchestras in London'), its membership consisted largely of ladies from aristocratic families, including the feminist activist Princess Sophia Duleep Singh. Entrance fee was originally 10 guineas and subscriptions were three guineas. The premises later boasted 70 bedrooms. Gentlemen were admitted only to the dining room, tea room and smoking lounge. The Club had a membership in 1901 of 4,000. It still existed in 1952, but had closed by 1956.

Enfranchisement of Women Committee

Founded in October–November 1866, its aim was to continue the work of the **London Provisional Petition Committee**. It ceased on the formation of the **London National Society for Women's Suffrage** in 1867.

(Crawford, 1999)

Enterprise Club

Founded in 1899 at 57–8 Leadenhall Street, London, the Club was intended as a social centre for women over 18 years of age engaged in clerical work. A registry of lady clerks was kept and there was a provident fund in connection with the Society. It later moved to 44 King Street, London, EC. In 1909 the secretary was Freda Wale.

Equal Compensation for War Injuries Committee

This committee of women MPs was formed on an initiative of the **British Federation of Business and Professional Women** and the **National Association of Women Civil Servants** in 1941 to campaign for the annulment of the provision for differential payments for men and women in the case of war injuries. The chairman was Mavis Tate; other committee members included Thelma Cazalet-Keir, Edith Summerskill and Megan Lloyd George. Caroline Haslett was prominent in extra-parliamentary campaigning. The Committee achieved its objective on 7 April 1943. An Advisory Council remained in being to oversee the application of compensation.

Equal Pay Campaign Committee

This developed from the experience of the **Equal Compensation for War Injuries Committee**. In 1944 Caroline Haslett asked Mavis Tate, MP, to sound out support among women MPs for a committee to campaign for equal pay between men and women. Although in fact the only women MPs who initially supported this initiative were Lady Astor and Edith Summerskill, a Committee was formed in 1944, consisting of representatives from the **Council of Women Civil Servants**, the **National Association of Women Civil Servants**, the Joint Committee of Women in the Civil Service, the **National Council of Women**, the **National Union of Women Teachers**, the **Association of Assistant Mistresses**, the **Women's Institute** and the **British Federation of Business and Professional Women**. From the outset it was apparent that the Committee's interest was exclusively in the white-collar occupations of teaching and the Civil Service, effectively excluding anything manual or outside public-service employment. This meant that it had little support from the major unions.

The campaign looked as though it was going to score an immediate success on 28 March 1944, when in the House of Commons Thelma Cazalet-Keir, MP, proposed an amendement to R. A. Butler's Education Bill, which was then going through its parliamentary stages, proposing equal pay for women teachers in the state system; the amendment was resisted by the government, but was narrowly adopted by the House. The triumph was extremely short-lived; the following day the government referred the amendment back, this time making the vote one of confidence in the government (the blame for this extraordinary tactic has traditionally be attributed to Churchill, but it seems more properly to rest with Ernest Bevin). As a consolation, the government set up a Royal Commission on Equal Pay, which reported in 1946. In fact, the campaign gathered strength through the late 1940s and early 1950s, despite the death in 1947 of Mavis Tate (her position as chairman was assumed by Thelma Cazalet-Keir). In 1950 the Committee even commissioned a short film, *To be a Woman*, from the director Jill Craigie. However, the main activity of the Committee consisted less in public propaganda than in lobbying such influential quarters as the Chancellor of the Exchequer and the Ministries of Labour and Education. In 1955 the lobbying paid off, when the government announced that equal pay in the Civil Service and in teaching would be phased in over the period 1956–61. Thereupon, the Committee decided that its aims had been met, and dissolved itself in 1955.

(Smith, 1992)

Equal Pay League *see* **National Federation of Women Teachers**

Equal Political Rights Campaign

This organisation was founded by Lady Rhondda in 1926 to unite all groups working for women's equal franchise. It ceased to exist in 1928.

(Crawford, 1999; Eoff, 1991)

Espérance Girls' Club

It began in about 1888 as a Club for Working Girls of the West London Mission on the initiative of Mary Neal, then a member of the Mission's **Sisters of the People**. In 1891 she was joined by Emmeline Pethick (later Pethick-Lawrence). A major part of their work was concerned with drama and dancing, and in 1895 they broke away from the Mission to found their own Espérance Girls' Club, which gave them more scope for experimenting with these activities. The Club also started a tailoring establishment, Maison Espérance, to provide better employment opportunities for young women in the West End, with better than average pay and conditions. The Club had an annual seaside outing or holiday, and in 1901 established the Green Lady Hostel at Littlehampton, Sussex. In 1905 the Club became explicitly involved with the folk song and dance revival, at first with the support of Cecil Sharp, who spoke at the Club's Christmas Party that year at the Passmore Edwards Settlement. In 1906, girls of the Espérance Club began going out across England teaching the revived folk-dances, especially Morris dancing, and by 1909 Mary Neal and the Club held a leading position in the folk-dance and song revival, although by then rivalry with Cecil Sharp had become open. In 1910 Mary Neal introduced a side of young men to dance alongside her girls, and changed the name of the Club to the Espérance Guild. This continued in rivalry with Sharp's English Folk Dance Society until the outbreak of war in August 1914, after which it appears to have ceased activity.

(Judge, 1989)

European Union of Women: British Section

This organisation was set up by the Conservative Party in 1953 to safeguard and protect the rights of individuals and encourage them to play an active part in political and public life in the United Kingdom and in Europe and in maintaining the cultural heritage of Europe. In 1997 there were between 3,000 and 4,000 members.

Evangelical Protestant Deaconesses' Institution

Established in 1868, the Institution's objective was to train educated women as nursing sisters in order to provide public hospitals and private families with skilled staff. By 1878 the Institution had 23 deaconesses and 10 probationers in Britain, as well as others tending to the sick and wounded Russian and Turkish soldiers in Bulgaria.

Fabian Women's Group

The Group first met in the drawing-room of Maud Pember Reeves in March 1908. Its aims were to further the principle of equal citizenship within and outside the Fabian Society, and to 'study women's economic independence in relation to socialism'. Prominent members included Beatrice Webb, Barbara Drake, Susan Lawrence, Margaret Bondfield, Charlotte Shaw, Edith Nesbitt and Alice Clark, as well as Maud Pember Reeves; the Group's first secretary was Charlotte Wilson. Members' activity included standing as candidates in local elections for Poor Law Guardians, working within the **Women's Labour League**, trade unions and other labour organisations, campaigning for divorce reform and prison reform, for child welfare and, from 1911, for women's suffrage. In 1910 it had 211 members, mainly well-to-do women. Its members were particularly active in writing books, articles and, above all, Fabian Tracts. It does not seem to have been active as a group after the outbreak of war in 1914, although many of its members continued to be prominent in other socialist feminist bodies.

(Alexander, 1987)

Factory Helpers' Union

The Union, based at 16a Old Cavendish Street, London, was established in 1886 by Lady Kinnaird under the auspices of the **Young Women's Christian Association** to promote the spiritual and moral welfare of factory and working girls. The Union based its work on eight departments: weekly visits by ladies to factories and laundries; establishment of evening homes and institutes; a total abstainers' union; a flower mission; a society for making clothes; a letter mission; homes for convalescents, the sick and the elderly; and receiving and distribution of cast-off clothing. A central committee oversaw the 61 missions, of which 19 were in the provinces and 42 in London.

Family Endowment Council

Set up by Eleanor Rathbone in 1917 as the Family Endowment Committee which changed its title to the Family Endowment Council the following year. It campaigned for a 12s 6d. weekly allowance for mothers, five shillings for the first child and 3s 6d. for subsequent children. The Society's object was 'to collect and disseminate information, to promote discussion and to take action, with a view to bringing about as quickly as possible a more adequate method than at present of making provision for families'. It hoped to put an end to the economic subjection of the wife and raise the status of married women. It was also argued that wages could then be paid on a bachelor/spinster basis, thus removing the main impediment to equal pay. The Council was a very active body. It distributed 30,000 pamphlets after the publication in 1926 of the Royal Commission on the Coal Industry which recommended allowances for miners' wives. The original Committee consisted of four feminists and three Socialists including Eleanor Rathbone, Elizabeth Macadam and Eva Hubback with Mary Stocks as one of the secretaries. Male presidents in the 1920s included Lord Balfour of Burleigh and William Beveridge. It became the Family Endowment Society in 1925 when the membership was only 77. After 1926 the original purpose of the Society, to campaign for allowances as payment to mothers for housework and as a means to equal pay, faded from view. When the Beveridge Committee was set up in the Second World War, the Society gave evidence to it. Family allowances were granted by Act of Parliament in 1945.

(Macnicol, 1980)

Family Planning Association

In the late 1930s, panics over depopulation had rendered the term 'birth control' problematical. This led the **National Birth Control Association** in May 1939 to change its name to the Family Planning Association. In doing so it effectively redefined itself; its new objectives were to provide facilities 'so that married people may space or limit their families and thus mitigate the evils of ill-health and poverty', and to establish women's centres which would give advice not only on contraception, but on infertility, on marriage difficulties and minor gynaecological ailments. The emphasis was thus shifted further towards enabling women to be in control of their own fertility. However, during the Second World War, its activity was reduced to a minimum, not least because two of its most vigorous leaders, Lady Denman and Margaret Pyke, had become deeply involved with the work of the **Women's Land Army**. The Association's headquarters was removed to Bournemouth for the duration. Many clinics, especially in London, suffered from bomb damage, and a shortage of rubber affected the supply of caps. In 1945 the Association submitted evidence to the Royal Commission on Population. However, when the National Health Service was formed in 1948, family planning provision was not included, and the Association judged that the time was not ripe to lobby for this; there were also fears that incorporation of clinics within a government scheme might undermine their autonomy.

In the post-war years it increasingly became involved in advice on marriage problems and sexuality in general, and had to confer with the Marriage Guidance Council in order to avoid duplication of effort. Much of the Association's efforts at this time were directed towards putting pressure on local health authorities to provide family-planning facilities; this met with varying success, depending on such factors as the financial situation of the authority, the level of local Catholic opposition, and the opposition of Medical Officers of Health. After Lady Denman's death in 1954, the leading figures in the organisation were Margaret Pyke, Margaret Howard and Helena Wright. A major breakthrough came in November 1955 when the then Minister of Health, Ian Macleod, officially visited the Association on the occasion of its silver jubilee, and praised its work.

Throughout the late 1950s the Association was opening new clinics at the rate of two a month; by 1960 it had established 276 branches which in that year alone saw 495,903 patients. Another breakthrough occurred in August 1959 when Mervyn Stockwood, Anglican Bishop of Southwark, broadcast an appeal on BBC Radio as the *Week's Good Cause*. In the 1960s the advent of the contraceptive pill dramatically changed perceptions of birth control; even so, it was not until 1974 that family planning became officially part of the National Health Service. For most of its existence to date, the Association was based at 23 Mortimer Street, London, W1.

(Leathard, 1980)

FANY *see* **First Aid Nursing Yeomanry**

Fawcett Library *see* **London and National Society for Women's Service**

Fawcett Society *see* **London and National Society for Women's Service**

Federated Council of Women's Suffrage Societies

This was formed in 1912 to 'determine a united policy and action which all the constitutional societies might adopt'. In fact, it seems to have been mainly aimed at the smaller specialist societies (professional and religious).

(A. J. R., 1913)

Federation of Business and Professional Women's Clubs *see* **National Federation of Business and Professional Women's Clubs**

Federation of Women Civil Servants

The Federation, formed in 1916, consisted of an amalgamation of the **Civil Service Typists' Association** and the **Federation of Women Clerks**, and represented all clerical women except Writing Assistants. Its objective was to achieve equal pay and to this end was willing to co-operate with the Civil Service men's unions. With the assimilation of women into the general grades of employment in the Service in 1920, it was weakened by the recruitment campaigns conducted by other organisations. In 1932, with declining numbers, it amalgamated with the **Association of Women Clerks and Secretaries**, taking the title the **National Association of Women Civil Servants**.

(Humphreys, 1958)

Federation of Women Clerks

As women Civil Servants were in separate and lower grades than men and with a separate establishment, there was a strong incentive to press for equal pay and conditions of work. In 1913 the different Civil Service women's associations came together to form a **Federation of Women Clerks**. In 1916, it became part of the **Federation of Women Civil Servants**.

Federation of Women Zionists of Great Britain and Ireland

It was established in 1918 to help the Yishuv (the Jewish community in Palestine) by raising funds, by education, and by lobbying for an independent Jewish state. Its status as a separate women's organisation seems not entirely to have pleased the Zionist Federation, although it was not officially disapproved of; its historian described the relationship as somewhere between a marriage undergoing difficulties and a divorce that was never made absolute. It later affiliated to the Women's International Zionist Organisation (WIZO). The first president was Mrs Rebecca Sieff, the chairman (from 1920) was Mrs Ethel Solomon, and the honorary secretary was Miss Olga Ginzburg. By 1940 it had a membership of approximately 16,000 in 160 affiliated societies. Its organ was the *Jewish Women's Review* (later the *FWZ Review*, and at the time of writing *Vision*). Its original address was 107 Gloucester Place, London, W1.

(Sherr, 1968)

Female Bible and Domestic Mission *see* **London Biblewomen and Nurses' Mission**

Female Chartists

This organisation was active in Aberdeen in 1841.

(Taylor, 1983)

Female Middle-Class Emigration Society

This, the first Society with the express purpose of encouraging the emigration of British women to countries of the Empire, was founded by Maria S. Rye and Jane Lewin in 1862, at 12 Portugal Street, London. It was funded by voluntary contributions, which enabled it to supplement the assisted passage of women emigrants. Initially it experienced difficulties both of selection of applicants for emigration and with their adaptation to their new circumstances; the term 'middle-class' was interpreted fairly broadly, and most of the early emigrants had few if any professional qualifications. Although it was regarded as being mainly for the benefit of 'the deserving governess', many of those helped were of lower social standing than this. The Society's work was taken over in 1884 by the **Colonial Emigration Society**.

(Hammerton, 1979; Monk, 1963)

Female Political Union

Established at Newcastle upon Tyne in about 1839, it consisted of Chartist women, acting as 'auxiliaries' to support the claims of working-class men for political representation.

(Taylor, 1983)

Female Radical Association

This organisation, supporting Chartists, was in existence at Carlisle in 1838.

(Taylor, 1983)

Female Radicals

Active at Ashton-under-Lyne, Lancashire, in 1839, this was a women's Chartists group.

(Taylor, 1983)

Female Servants' Home Society

Established in 1837 with the subtitle 'for the encouragement of faithful female servants', it served to provide a safe home for female servants of good character who were out of work, and recommended servants who remained in the same situation for 12 months or even longer. It provided for the needs of the 10,000 female domestics in London, especially those who were in transition between posts. In 1861 it had four homes. A register of servants who were available was kept at each home for the use of potential employers. In 1887, some 1,225 young women were accommodated. Its patron was the Duchess of Cambridge.

Feminist League

This elusive organisation appears to have been active in the 1920s. Its founder was Alice Abadam.

First Aid Nursing Yeomanry

Founded in 1907 by Edward Charles Baker, a veteran of the Sudan campaign, the Corps was to assist, in the event of war, in tending the wounded on the battlefield. Recruits were to be qualified in first aid and home nursing and were required to pass courses in horsemanship, veterinary work, signalling and camp cookery. Women had to be aged between 17 and 35 and pay an enrolment fee and monthly subscriptions to the riding school, as they would act as mounted auxiliary to the Royal Army Medical Corps. Drilling and training in camps was part of the routine. In September 1914, shortly after the First World War broke out, a unit of the FANY Corps went to Belgium and also to Calais to staff hospitals there. Suspicion on the part of the Army Medical Service and the independent nature of the Corps itself led to friction. As a result, the main work of the FANY changed: it provided drivers for ambulance convoys for the Belgium, French and British armies, working often under difficult and dangerous conditions.

After the War, when the **Women's Army Auxiliary Corps** and the **Women's Royal Naval Service** were disbanded, the FANY remained. It drove cars and lorries during the 1926 General Strike and was officially recognised by the War Office in 1927. In 1935, it was given a new title, the Women's Transport Service, to reflect the changed nature of its work. Although the Corps lost a number of its members in 1938 when they were invited to form the Motor Driver Companies of the newly formed **Auxiliary Territorial Service**, nevertheless it continued during the Second World War to provide valuable support to the effort. In 1939, members served in Finland when it was invaded by Russia. A Polish Unit was attached to Polish Forces in England and the FANYs saw service in the Middle East, East Africa, India and Ceylon. Of the 39 members who were recruited for the Special Operations Executive, 13 perished. Three FANYS were awarded the George Cross.

(Popham, 1984)

Flower Girls' Mission

Although the Flower Girls' Mission was established by a man, Mr Groom, in 1886 it soon became an all-women's organisation. It gave classes for flower sellers in reading and sewing at a house near Covent Garden, opened a penny bank and later a free evening school. This was not the first initiative in this field. In 1879 a Flower Girls' Brigade was started by Angela Burdett-Coutts for flower-sellers between the ages of 13 and 15. Special stands were arranged by the Mission so that the girls could carry on their trade in safety and a nearby shelter provided sewing and cookery classes. Burdett-Coutts also provided regular employment for the girls by opening an artificial flower factory. Their products became well-known and were sold all over the world. Later, a Flower Girls' Guild was formed under the patronage of Princess Alexandra for 'sound and sober girls'. Rest rooms, evening classes and entertainments were also provided. An Emily Loan Fund, instituted in memory of Emily, Countess of Shaftesbury, was available for loans of potato ovens, coffee stands and other equipment in winter.

Forum Club

It was founded in 1919 as the London Centre for **Women's Institute** members. Its facilities were extensive. The Club had, apart from the usual premises, rooms set aside for bridge, billiards, photography, a hairdressing salon and a picture exhibition gallery. There was also accommodation for members' maids. Prominent members included Sybil, Lady Rhondda (mother of Margaret, Lady Rhondda, founder of *Time and Tide*) and the playwright and actress Elizabeth Robins. The Club was still active in 1934.

Forward Cymric Suffrage Union for Men and Women

In 1911 Welsh suffragists felt the need to organise separately, and Mrs Edith Mansell-Moullin established a Cymric Suffrage Union, with Mrs D. A. Thomas, mother of the future Lady Rhondda, as president.

However, because of internal disagreement over reactions to the government's proposed manhood suffrage bill, Mrs Mansell Moullin re-established the organisation in 1912 as the Forward Cymric Suffrage Union, with a strongly militant agenda. Its motto was the Eisteddfodic 'O Iesu, na ad gamwaith' (Jesus, do not allow injustice). As well as branches in Wales, it included a London branch (in fact the officers were all London-based). It was later associated with the East London Federation of Suffragettes, when the latter broke away from the **Women's Social and Political Union**. It appears to have ceased activity by 1916.

(A. J. R., 1913; Crawford, 1999)

Forward Suffrage Union

This was formed in 1907 by Eva McLaren, Marie Corbett and Frances Maxwell to concentrate the suffrage activities of Liberal women affiliated to the **Women's Liberal Federation**, in the spirit of the **Union of Practical Suffragists**. It was still in existence in 1913.

(A. J. R., 1913; Crawford, 1999)

Free Church League for Women's Suffrage

Founded in 1910 by Miss L. E. Turquand, Mrs Strickland and the Rev. Hatty Baker, this Nonconformist body tended to the evangelical, but also to the ecumenical. From 1913 it had its own paper, which ran until 1920 (from 1916 to 1920 it was called the *Coming Day*).

(A. J. R., 1913; Crawford, 1999)

Freedom of Labour Defence

After the failure of the **Women's Employment Defence League**, this organisation was established in October 1899 by Jessie Boucherett, a staunch Conservative and a keen individualist as well as a prominent feminist, together with her close associate, Helen Blackburn, who was eager to promote the wider employment of women. The aim of the FLD was to encourage self-help for women workers and oppose restrictive factory legislation (otherwise known as protective legislation). As the chairman of the Central Committee, William Thompson Sharpe, a Conservative MP, commented, 'Grown up men and women should be perfectly free to do what they like with their own: surely the labour of their own bodies is their own property, and ought to be left to them.' The only members belonging to it who were not home workers were the chairman, the vice-chairman and the treasurer. Its first secretary was Miss Vynne who was succeeded by Miss Alice Jackson in 1906. Meetings were held once a fortnight at the Cadogan Club. The views of the FLD were disseminated through leaflets, public meetings, letters to the press and memorials and petitions to the Home Secretary. In 1901, the organisation opposed no fewer than four Bills then before the House of Commons – the Factory and Workshops Act Amendment Bill, the Home Industries Bill, the Shops Bill and the Shops (Early Closing) Bill. The 1902 Employment of Children Act was also opposed on the grounds that 'a different kind of work is a relaxation from books, and when, as is frequently the case, it is outdoor work, it is very good for the health'. Unregulated home work was encouraged. By the middle of the first decade of the twentieth century, the FLD, never a popular cause, had lost much of its support. The campaign against restrictive legislation was taken up later from a very different standpoint by the **Open Door Council**.

Friends' League for Women's Suffrage

Formed probably in 1912 by Alice Clark and Guilema Crosfield to organise Quakers to campaign for women's suffrage, in 1913 it had 19 branches. It was still in existence in 1917.

(A. J. R., 1913; Crawford, 1999)

Friends of Women's Suffrage

This was a supporters' body organised in 1912 by the **National Union of Women's Suffrage Societies** for those who did not wish, or could not afford, to be full members. By mid-1913 it had enrolled 46,662 members. It caused some controversy within the NUWSS, as did the Election Fighting Fund, as being too Labour-oriented. It does not appear to have survived the outbreak of the First World War.

(Crawford, 1999)

Gentlewomen's Cigarette Employment Society

The Society was formed to provide remunerative work for ladies who needed to earn a living. Only the best tobacco was used and the finished products were sold at the same price as other retailers. Pay was five shillings per 1,000 cigarettes, and represented one day's fair work. It shared the same premises as the **Lady Guides' Association**, which by 1891 had moved to 352 Strand, London and was organised by Miss Edith A. Davis, founder of the Lady Guides' Association and Mrs Arthur Abbott, the latter an ex-member of the Lady Guides' Association.

Gentlewomen's Employment Club

In 1885, Adele Younghusband contacted Louisa M. Hubbard, editor of the *Englishwoman's Year Book* and *Work and Leisure,* and, inspired especially by the latter publication, took over the task of interviewing all new applicants who were seeking employment. She acted as a free registry, putting applicants in touch with potential employers. Miss Younghusband also organised a workroom and instruction for ladies who wished to become dressmakers. In 1889, Louisa Hubbard provided sufficient rent for Miss Younghusband to open a Dressmaking, Millinery and Plain Needlework Club at 7c Lower Belgrave Street, London, SW. Its title was changed the following year to the Gentlewomen's Employment Club. It provided practical training for ladies wishing to be trained for such callings as laundry matrons, visiting dressmakers and nursery superintendents as well as governesses and companions. Accommodation was provided for a limited number of working ladies whilst training. It later moved to 155 Victoria Street, London, SW. It was still active in 1909.

Girl Covenanters

An interdenominational youth movement with an Evangelical basis, its objective was to prevent the falling away of girls from Sunday School after the age of 11. Membership was open from this age and members were entitled to wear the Girl Covenanter badge after attending Bible Class on ten consecutive occasions. Recreational activity was encouraged, with camps, sports rallies and netball and hockey tournaments being mounted.

Girl Crusaders' Union

The Union's objective was to train its members for Christian (Evangelical) leadership, which included missionary and church work. Rallies and camps were annual features, the camps attracting more than 1,000 girls each year. The Union became an international youth movement, with classes in China, India, Africa and Switzerland.

Girl Guides *see* Guide Association

Girls' Friendly Society

Launched by Mrs Mary Townsend at a meeting of Anglican ladies at Lambeth Palace in May 1874 to meet the needs mainly of girls in domestic service, its objectives were twofold: 'to bind together in one Society ladies as associates and working girls and young men as members for mutual help (religious and secular), for sympathy and for prayer' and 'to encourage purity of life, dutifulness to parents, faithfulness to employers and thrift'. (In fact the Society became the province exclusively of women.) Members had to be unmarried and no girl who had not a 'virtuous character' was admitted. Membership, which was not confined exclusively to the Church of England, was strongest in country areas, and the Society prepared young women for some of the dangers arising out of moving to towns and cities. Within six years of its founding the Society had recruited 47,000 members. Like its sister organisation, the **Young Women's Christian Association**, it provided lodging houses, training schemes and weekly sewing classes, and mounted garden fêtes. It also provided cheap and wholesome literature by means of circulating libraries and magazines.

As Brian Harrison has pointed out, the GFS, by promoting deference through recreation, evolved Conservative techniques which became better known in the **Primrose League** and trained many of its workers. Branches were run by Anglican associates with lady presidents for each diocese. The secretaries of all the branches in a diocese formed a diocesan council. The presidents of the diocesan councils, together with the Colonial and foreign presidents, heads of departments and 10 elected members, formed the Central Council. Its organs were *Friendly Leaves*, *Friendly Work* and the monthly *GFS Associates' Journal*, later the *GFSA Review*; in addition, 'useful literature' was distributed to all branches. The GFS made great efforts to improve the status of domestic servants by establishing registry offices which found positions for 4,000 members a year. The Society was established in Ireland in 1877. The movement spread rapidly, to the Continent, the United States of America and Russia. The Society opened a small emigration department in 1883 and a Colonial Committee was started in 1896. In 1913 at the height of its popularity, the Society had 39,926 associates and 197,493 members in England and Wales. The Society, which still flourishes, has its headquarters at 126 Queen's Gate, London, SW7.

(Harrison, 1973; Heath-Stubbs, 1935)

Girls' Guildry

The objects of the Girls' Guildry, which was founded in Glasgow in 1900 by Dr William Francis Somerville, a Free Church minister, were to 'help girls to be followers of the Lord Jesus Christ, to promote girls' discipline, self-respect, helpfulness and reverence, the ultimate aim being full membership of the Christian church and wise and responsible Christian Worship'. Somerville was concerned by the lack of provision for girls after leaving Sunday School and he modelled the Girls' Guildry closely on that of the Boys' Brigade. There were four groups within the Guildry: the Service Company for those aged 14 and upwards, the Guildry Company for ages 11 to 13, the Greenwood for Speedwells for 8 to10 year olds, and the Garden for children aged 5 to 7. A wide range of activities was offered; keep-fit classes, drill, Bible study, handicrafts and needlework, first aid and nursing, citizenship, cooking, rambling and camping. Members in each group attended public worship, Bible class or

Sunday School. By 1939, membership had reached 24,000. In 1965, it joined with the **Girls' Life Brigade** and the Girls' Brigade of Ireland to form the Girls' Brigade.

Girls' Letter Guild

The object of the Guild, begun in Birmingham by Miss Knowles in November 1889, was to bring 'cultured Christian ladies' into contact with factory and mill girls. Each lady wrote to one or more girls once a month on practical topics such as dress, health and amusements. The Guild stated that as a result the girls willingly confided their troubles and difficulties to the ladies, who became a good influence over them. The movement spread to other parts of the country and domestics and shop girls became eligible to join the scheme.

Girls' Life Brigade

The Girls' Life Brigade was established in 1902 and was closely modelled on the Boys' Life Brigade. The latter had been started in Glasgow in the 1880s by Dr William Francis Somerville, who later formed the **Girls' Guildry**. The Girls' Life Brigade promoted temperance on the lines of the National Sunday School Union and much of its activity initially was based on military lines. It had been preceded by the founding of a Girls' Brigade in Dublin in 1893, which provided weekday evening activities for Sunday School girls in Irish Protestant churches. The movement became popular in England and by 1914 had recruited 4,000 members. In 1965, it was amalgamated with the Girls' Guildry and the Girls' Brigade of Ireland to form the Girls' Brigade. The new organisation had 180,000 members at home and overseas.

(Springhall, Fraser and Hoare, 1983)

Girls' Public Day School Trust

In 1871 the National Union for the Improvement of the Education of Women of All Classes, later the **Women's Education Union**, was established by Maria Grey as part of her campaign to improve women's secondary education. Together with her sister, Emily Shirreff, she drew up a scheme for building public day schools which would charge reasonable fees and be self-supporting for middle-class girls in West London. Finance was to be raised by shareholders, having limited liability, in the Girls' Public Day School Company (later Trust) which was incorporated in June 1872. The prospectus stated that its object was 'to establish and maintain in such parts of London and the provinces as from time to time decided on superior day schools, at a moderate cost, for girls of all classes as above those provided for by the Elementary Education Act'. The first two were

Durham House, Norland Square, Chelsea and Notting Hill High School, both opened in 1873. The venture was successful and the Company's schools were subsequently established throughout the country. Starting in 1873 with 20 students, by 1879 there were 17 schools and a total of 2,804 pupils. By 1901, 38 schools had been founded. In 1904, the Company was converted into a Trust and its title was amended. It is still active at the time of writing. It is now named the Girls' Day School Trust.

(Magnus, 1923)

Girls' Union Club

Its headquarters were at the **Soho Club**, 59 Greek Street, London, which attempted to link all clubs, and evening and recreation rooms for girls and young women in England. It published a monthly magazine. By the end of 1886, almost 30 clubs and evening classes had joined the Union.

Glasgow and West of Scotland Society for Women's Suffrage

Formed in 1870, in 1872 it associated itself with the Central Committee of the **National Society of Women's Suffrage**. Its political affiliation was strongly Liberal, which promoted it in 1912 to oppose the Election Fighting Fund of the **National Union of Women's Suffrage Societies**. It ceased activity on the outbreak of war in August 1914.

(Crawford, 1999)

Glasgow Society of Lady Artists' Club

This Club, for the use of members of the Glasgow Society of Lady Artists, was formed in 1882 with premises at 5 Blythswood Square, Glasgow, with the aim of encouraging art, music and literature and to provide a convenient meeting place. Residential accommodation was also available. Subscriptions were three guineas for lay members and one guinea for artists. Annual exhibitions of members' work were held. The house itself, built early in the nineteenth century, was strikingly altered both internally and externally by Charles Rennie Mackintosh in 1908.

Governesses' Benevolent Institution

Established in 1841 initially 'to afford relief privately and delicately to ladies in temporary distress' by providing small sums of money to needy governesses, the Institution soon opened a home for elderly governesses and another for those temporarily out of work. A more ambitious object was to train governesses to become teachers to a high standard in order to command a reasonable salary. Frederick Denison Maurice and others provided lectures, and a house was taken in Harley Street, London, next to the Governesses' Benevolent Institution after one of the Queen's ladies-in-waiting, Amelia Murray, had collected sufficient money for this purpose. Queen's College for Women opened in 1848. Courses in the first instance were primarily for the benefit of governesses but they were also open to 'ladies' over 12 years of age. Original students included Frances Mary Buss and Dorothea Beale.

Grand Lodge of Miscellaneous Female Operatives

This women's trade union was based in the Charlotte Street Institution, London, in 1834.

(Taylor, 1983)

Green Park Club

Founded in 1891 at 10 Grafton Street, Piccadilly, like the **Alexandra Club**, its membership was limited to ladies eligible to attend the Queen's Drawing Rooms. Its chief feature was its musical and dramatic entertainments, given on alternate Fridays, which attracted celebrities such as Nellie Melba, Maude Valérie White, Sarah Bernhardt and Beerbohm Tree. In 1901, when its president was Lady Edward Spencer Churchill, it had 500 members.

Grosvenor Crescent Club

Established in 1897 at 15 Grosvenor Crescent, London by Leonora Wynford Philipps as an offshoot of the **Pioneer Club**, its activities were similar to those of its parent organisation. It was a social club with residential accommodation and included amongst its members Mrs Henry Sidgwick and the Duchess of Aberdeen. The Club offered a low rate of subscription to professional women. Its object was to provide a centre for women interested in literature, art, science, philanthropic work and social questions of the day. Its special feature was the lively debates held there. Although its origin was formally the result of a split within the **Pioneer Club**, relations between the two clubs seemed to be friendly, and many women were members of both. Its support of women's self-improvement was expressed particularly by the fact that it gave initial accommodation to the **Women's Institute**. In 1909 membership was five guineas, and the entrance fee five guineas. It seems still to have been active in 1913.

Guide Association

Robert Baden-Powell, a soldier and national hero for his part in withstanding the siege of Mafeking during the Boer War, had developed scouting when he was in India as part of the training of soldiers. Badges were awarded for efficiency and the men were encouraged to develop initiative. His scheme was first set out in his book *Aids to Scouting* (1899).

Baden-Powell held a trial camp for boys from different social classes in 1907, encouraging them to work in small units and to high standards of accomplishment and behaviour. This led to the publication of *Scouting for Boys* (1908) and the establishment of Boy Scouts troops throughout the country. Baden-Powell became Chief Scout. It was at the Scouts' first large rally at the Crystal Palace in September 1909 that a group of 'Girl Scouts' invaded the proceedings and insisted on being inspected by Baden-Powell. His enthusiasm at first for a parallel organisation to the Boy Scouts was only lukewarm, but with his sister Agnes, he wrote a manual called *Girl Guides: a Suggestion for Character Training for Girls.* In May 1910 the Girl Guides' Association was formed with Agnes as president and Princess Louise, Duchess of Argyll, as its first Royal patron. In a short space of time some 8,000 girls joined the movement.

The ten Code Laws embraced a practical ethical code, which comprised honour, loyalty, helpfulness, friendliness, courtesy, kindness to animals, obedience, cheerfulness, thrift and clean living. The religious basis was emphasised and girls were encouraged to become full and active members of their own church. In 1914 the Brownies, a junior version of the Guides, came into existence. Both groups worked for different badges, as is still the case, ranging from cooking to interpreting. In 1918, Baden-Powell's young wife, Olave, became Chief Guide, and 12 years later, World Chief Guide, an indication of the movement's international appeal. The Girl Guides' Association received its Royal Charter of incorporation in 1922; children of the Royal family have traditionally been associated with the movement. World-wide links have led the Guides to support a number of initiatives in countries overseas, such as community projects, literacy, health and building. In 1987, a new section was formed, the Rainbow Guides, for girls younger than Brownies. The six age categories now are Rainbow Guide, 5–7 years, Brownie Guide, 7–10, Guide, 10–14, Ranger Guide, 14–25, Young Leader, 14–18 and Adult Leader, 18–65. One in three girls aged 5–10 is a member of the Association, and there are over 60,000 adult leaders. The Girl Guides' Association became the Guide Association in 1994.

(Springhall, 1977; Thompson, 1990)

Guild of Women Binders

This was formed in 1898 by the bookseller Frank Karslake, and it worked in conjunction with Karslakes's Hampstead Bindery. Its origin derived from the Victorian Era Exhibition of 1897, when Karslake noticed the work of women binders displayed there and persuaded the women to take part in an exhibition of women's binding. He then acted as agent for many of them, especially Annie MacDonald, Jessie MacGibbon, Mrs Douglas MacLagan,

Jean Pagin and Phoebe Traquair. The Guild was based at 61 Charing Cross Road, London, W. It produced a number of publications, notably *Bindings of Tomorrow* (1902). The Guild appears to have ceased by 1904.

(Tidcombe, 1996)

Gymnastic Teachers' Suffrage Society

This was formed in 1908 to organise gymnastic teachers for suffrage, and 'to protect the profession from restrictive legislation'. Its president was Mary Hankinson, and its treasurer was Miss May King. It appears to have ceased activity in 1915.

(A. J. R., 1913; Crawford, 1999)

Halcyon Club

This Club, situated at 13–14 Cork Street, Burlington Gardens, London, was open to women who held degrees in medicine, science and the arts or were engaged, either professionally or as 'serious amateurs', in art, music, literature, handicrafts, horticulture and other occupations. Concerts, lectures and exhibitions of members' work were frequently arranged. Every member was required to hold at least one share in the Club in order to provide working capital. Prominent members included Maude Royden and Beatrice Harraden. The Club was still in existence in the early 1920s.

Health Visitors' Association

In 1896 the first seven women sanitary inspectors based in London founded the Women's Sanitary Inspectors' Association. Ten years later there were 63 members and membership was extended to the provinces. Its aims were to safeguard the interests and improve the status of women public-health workers and to promote the interchange of relevant technical and professional knowledge. There were two grades of membership: full membership, those holding certificates recognised by the Local Government Board and who were employed by a sanitary authority, and associate membership, those qualified but not so employed. Meetings were held the first Wednesday in the month from October to May at the Royal Sanitary Institute, 90 Buckingham Palace Road, London, SW1. In 1908, the London County Council required that health visitors should undertake a two-year course of training after appointment. The name of the organisation was changed in 1915 to Women's Sanitary Inspectors and Health Visitors' Association. By 1918 the Association consisted of 3,000 health visitors, school nurses, TB visitors, midwives and matrons of day nurseries. The Association again changed its name in 1929 to the Women's Public Health Visitors' Association. In 1918 it had affiliated to the **National Union of Women Workers**, and in 1924 it had been the first health service union to affiliate to the Trades Union Congress. Midwifery was to remain an essential prerequisite for health visitors until 1938. After the Second World War health visitors' activities were extended by the School Health Service Act and National Health Service Act. The latter Act extended their work to focus on the health of the whole family as well as aftercare following hospitalisation, and long-term illness. In 1962, with the majority of its members being health visitors, it became the Health Visitors' Association.

(Allan and Allan, 1982)

Healthy and Artistic Dress Union

The Union was probably founded in 1901, with Henry Holliday as president; vice-presidents included Louise Jopling, Baroness von Hügel, Mrs G. F. Watts and Walter Crane. Committee members included Janet Ashbee, Sophie Bryant, and Mrs C. F. Andrews. The secretary was Maude A. Biggs, and the treasurer Amy G. Theobald. The Union's objects were 'to promote such improvement in dress as shall be consistent with health, comfort, activity and graceful appearance, yet shall not involve any obvious departure from recognised conventional modes'. Although theoretically aimed at both sexes, it was overwhelmingly focused on women's clothing. Annual subscription was four shillings, which included subscription to the Union's organ, the *Dress Review.* It does not appear to have survived beyond 1906.

Hibernian Ladies' Prison Association

In 1829, Elizabeth Fry visited Dublin to co-ordinate existing Ladies' Visiting Associations and to encourage the creation of new ones. Numbers increased for a time but the enthusiasm faded away after the 1830s.

Honor Club

Located in Fitzroy Square, London, it catered for 'better class working girls of the West End'. A different lady visitor superintended the club every night. A full programme was arranged for the young women. On Monday night, dancing was arranged and members could consult a women doctor if necessary, Tuesday night was gymnasium night, Wednesday was devoted to games, and Thursday was devoted to embroidery and poetry. A special Honor brooch was given to girls who had been members for over three years. Subscription rates were sixpence a month for 14 year olds and 1/6d for those over 17.

Honourable Fraternity of Ancient Freemasons

It was founded on 27 November 1913 by Mrs J. Elizabeth Boswell-Reid, who became Grand Master. Her Deputy was Mrs Seton Challen, her daughter, who succeeded her as Grand Master on her death in 1933. It claimed to be the only Masonic Fraternity organised entirely by women for women; in 1934 it had six lodges. Its organ was the *Ray*, succeeded by the *Woman Freemason.* An interesting point is that it used the conventional male terminology; for instance, not only were the officers known as 'Masters', but fellow women Masons were always addressed and referred to as 'Brethren'. In 1934 its address was The Temple, St Ermin's, Westminster, London, SW1.

Ilchester Club

The address of this Club was Ilchester Gardens, Hyde Park, London and was founded in 1897 for women of limited means. It had a smoking room, dining room, drawing room, music room and a silent room. Residential accommodation was provided, with 40 bedrooms.

Imperial Residential Club

The Club was for professional gentlewomen and was situated at 4–6 Lexham Gardens, Cromwell Road, South Kensington, London, in 1906. The honorary secretary was Miss H. J. Russell.

Independent National Union of Women Boot and Shoe Workers

By 1908 the National Union of Boot and Shoe Rivetters, founded in 1904, had failed to negotiate a fixed minimum rate for women employed in the trade. At its annual conference that year, it was agreed to implement this change, but no action followed. Two years previously, the Leicester Women's Branch, with over a 1,000 members, had appointed as its secretary Lizzie Watson, a lively and skilful organiser. She asked Margaret Bondfield to negotiate on behalf of the women and threatened to form a breakaway union. She was suspended by the National Union for her activities and established in 1910 the National Independent Union of Women Boot and Shoe Workers, campaigning strongly against the men's union. Six hundred of the Leicester Women's Branch joined the new union. It negotiated new rates for members, giving benefits to women up to the age of 50. From 1914 the National Union of Boot and Shoe Operatives agreed upon a national settlement to fix the minimum rates for women in the industry. By 1920 there were 1,200 members. The Independent National Union survived until the mid-1930s.

(Boston, 1980)

Independent Women's Social and Political Union *see* **Women's Social and Political Union**

International Women's Franchise Club

Established at 66 Russell Square, London, in February 1910, the Club later moved to 9 Grafton Street, Piccadilly. Its purpose was to provide a meeting place for men and women from all countries involved in the suffrage movement. It proved popular and soon had a membership of over 800; this had increased to 2,000 by July 1917. It had a feminist library from which members could borrow and a restaurant which provided meals at reasonable prices. The subscription rate was one guinea per annum, with Irish, Scottish and foreign members at 10/6d. Facing financial difficulties, the Club changed its name to the International Franchise Club in 1923 and closed in the following year.

Interval Club

This Club, which in 1938 was housed at 23 Dean Street, London, W1, was mainly to cater for Catholic members of the theatrical and allied professions, but it was also open to non-professionals. It had a residential house for ladies in nearby Soho Square. The honorary secretary was Miss Molly Balvaira Hewett.

Irish Home Industries Association

Founded by Lady Aberdeen in 1886 when her husband was Lord-Lieutenant of Ireland, the Irish Home Industries Association was an attempt to encourage knitting, sewing and weaving among the peasantry. They were supervised by 'residential ladies' and the products were sent for sale in two shops organised by the Association in Dublin and London.

(Luddy, 1995)

Irish Women Workers' Union

When James Larkin set up the Irish Transport and General Workers' Union in 1908 and became its general secretary, women were excluded from membership. Larkin's sister Delia subsequently established the Irish Women Workers' Union in September 1911. Suprisingly, James Larkin was its first president and Delia its secretary. Equally strange was the fact that Larkin set up a women's Irish Textile Workers' branch within his own union. In 1913 when the Irish Transport and General Workers' Unions was fighting for recognition which resulted in a lockout, the Irish Women Workers' Union aided the strikers and their families and followed with industrial action on their own account. Similarly, the Irish and Transport and General Workers' Union supported women in their efforts to obtain industrial and political rights and to raise their economic status.

Irish Women's Franchise League

The League was founded in Dublin on 4 November 1908 because of dissatisfaction with the methods of the **Irish Women's Suffrage and Local Government Association**. Originally, the secretary was Hanna Sheehy-Skeffington, and the treasurer Margaret E. Cousins. Later the secretaryship passed to Maud Joynt, then to Kathleen Emerson, and then to Margaret E. Cousins. It also had male members, prominent among them Francis Sheehy-Skeffington, James H. Cousins and Cruise O'Brien. Its tactics included militancy. In 1913 it was based at Ancient Concert Buildings, Great Brunswick Street, Dublin. From 1912, its organ was the *Irish Citizen*. Initally it kept a neutral stance on Home Rule, but it opposed the 1914–18 War, and after the killing of Francis Sheehy-Skeffington at the time of the Easter Rising in 1916 it supported Sinn Fein. From 1918, Sheehy-Skeffington was the League's president. In 1921 it opposed the Treaty. It does not appear to have continued beyond 1921.

(Levenson and Natterstad, 1986)

Irish Women's Suffrage and Local Government Association

This was started as the Dublin Women's Suffrage Association in 1874, when the secretaries were Anna M. Haslam and Miss McDowell. It held meetings, and sent petitions to the House of Commons in London from various parts of Ireland. In addition to the parliamentary suffrage it was vigorous in advancing women's position in local government, achieving the appointment of Ireland's first two women Poor Law Guardians. In 1898, its name changed to the Irish Women's Suffrage and Local Government Association. In 1913, Mrs Haslam was still its secretary at 163 Rathgar Road, Dublin, and the treasurer was Mrs Sanderson of 7 Mount Temple Terrace, Dartry Road, Co. Dublin. It had branches in Cork, Limerick and Mallow. The Association does not seem to have survived the War and the Easter Rising.

(A. J. R., 1913)

Irish Women's Suffrage Federation

This Federation was founded in the autumn of 1911 to unite scattered suffrage societies in Ireland in co-operation, while guaranteeing local autonomy. In 1913 its constituent societies included the Belfast Women's Suffrage Society (president, Miss Montgomery, secretary, Miss Walkington), the Irishwomen's Reform League (secretary, Ella Inglis), the Munster Women's Franchise Association, based at Cork, with branches in Waterford, Queenstown, Bandon, Skibbereen and Michelstown (secretary, Edith Œ. Somerville), the Lisburn Women's Suffrage Society (secretary, Mrs R. H. Metge); and societies at Warrenpoint, Armagh, Antrim, Kerry and Nenagh. The Federation was non-party, and as a whole non-militant, although it left individual members free to engage in militant acts if they so desired. It saw its function mainly as co-ordinating and lobbying, and it seems to have maintained good relations, at least up to 1914, with other Irish suffrage bodies, notably the **Irish Women's Franchise League**. In 1913, the Federation's president was the Hon. Mrs Spring-Rice, the secretaries were Louise Bennett and Miss Chevenix, the treasurer was Miss Moser, and committee members included Rosa Barrett and Margaret MacSwiney. Its headquarters was at 29 South Anne Street, Dublin. The badge bore its name in silver letters on a small Celtic cross in St Patrick's blue.

(A. J. R., 1913)

Irish Women's Suffrage Society

In 1913, this Society was based at 27 Donegall Place, Belfast; its secretary was Miss Robinson. It had a Derry (*sic*) branch, the secretary of which was Miss R. G. Gamble, Elagh Hall, Londonderry.

Jewish Association for the Protection of Girls and Women

Established in 1885, the Association (originally the Ladies' Association for Prevention and Rescue Work) under its president, Lady Rothschild, and its honorary secretary, Lady Battersea, carried out rescue work and had a home in Shepherd's Bush, London, for unmarried mothers and those rescued from prostitution, as well as a lodging house in the East End. The Association paid particular attention to the plight of friendless but respectable foreign girls who might 'fall prey to the unscrupulous on arrival at the docks'. It later changed its name to the Association for the Protection of Girls and Boys.

(Burman, 1996)

Jewish Girls' Brigade

In 1895, the Society of Maccabeans, an association of Jewish professional men and others, founded the Jewish Lads' Brigade 'to instil into a rising generation, from its earliest youth, habits of orderliness, cleanliness and honour, so that in learning to respect themselves, they would do credit to their community and country'. It modelled itself on the Church Lads' Brigade, and had a strong military flavour with an emphasis on drill, physical training, games and boxing. It was not until 1963 that the Jewish Girls' Brigade was formed on a similar pattern to that of the Boys'. The first Company, with a uniform of navy blue, was established in Liverpool and the movement then spread to other parts of the country. Activities included first aid, basketry, cooking and handicrafts as well as camping.

In 1974, with growing links between the two bodies, the name was changed to the Jewish Lads' and Girls' Brigade.
(Bunt, 1975; Kadish, 1996)

Jewish Ladies' Benevolent Loan Society

This Society, operating from 8 Duke Street, Aldgate, London, EC, was founded in 1844 with the purpose of granting loans of from one to ten pounds without interest or other charges to the deserving Jewish poor. Its president was Louisa, Lady de Rothschild, and admission to the Society was by nomination of the Committee. In 1885–86 its income was £2,572 4s 2d.

Jewish Ladies' Clothing Club

Established in November 1892 with its headquarters at 23 Oxford Square, Hyde Park, London, the Association's main objective was to supply articles of clothing to schools, charitable institutions and visiting committees for distribution among the needy. Its president was Mrs Leo de Rothschild and its honorary secretary, an American, Mrs Meyer A. Spielman.

Jewish League for Woman Suffrage

The League was formed in November 1912 'to unite Jewish suffragists of all shades of opinion, and … many would join a Jewish League where, otherwise, they would hesitate to join a purely political society'. It co-operated with the other religious suffrage societies. The minimum subscription was set at 3d, so as to enable Jewish suffragists of all classes to join. The Executive Committee included Inez Bensusan and Caroline Franklin. It appears to have ceased activity after 1918.
(A. J. R., 1913; Crawford, 1999)

Jewish Tailoresses' Union

Following the failure of male Jewish tailors to organise themselves, the London Jewish tailoresses formed their own union in 1903. It developed as a branch of the Amalgamated Society of Tailors. Dues were 3d. a week and after a year, members were eligible for strike pay, sick benefits and unemployment benefit.
(Black, 1988)

Josephine Butler Society *see* Association for Moral and Social Hygiene

Kelvin Club

It was founded in 1898 at 42 George Square, Glasgow. Candidates had to be nominated by two members and election was by a general ballot. Subscription was £2 12s 6d. and the entrance fee was three guineas. Wives and daughters of officers holding commissions in the Army and Navy were admitted as temporary members without entrance fee. In 1901 the president was Mrs J. G. A. Baird and the secretary Miss Mary Stewart Wright. There were 400 members. It was still active in 1909.

Kensington Ladies' Club

This Club, established in September 1893, at Granville Mansions, Earls Court, London, SW, was intended to provide a convenient home for ladies on moderate terms. Accommodation was available on both a permanent and temporary basis. Qualification for entry was 'gentlewomen by birth or position'. The entrance fee was 10s 6d. and the subscription one guinea annually. In 1904 there were 200 members. The president was Mrs A. D. Gill and the secretary Miss Faulkner.

Kensington Society

This Society was formed in 1865 to discuss informally the emancipation of women and related topics. The Society's name derived from the address of the secretary, Mrs Charlotte Manning, 44 Phillimore Gardens, Kensington, London,W8. Its secretary was Emily Davies, and members included Barbara Bodichon, Jessie Boucherett, Isa Craig, Elizabeth Garrett (Anderson), Helen Taylor and Elizabeth Wolstenholme (Elmy). Although to some extent superseded by the rise of the new suffrage societies, it continued until March 1868.

(Crawford, 1999)

Ladies' Army and Navy Club

Founded by Mr G. A. Dundas in 1909, the Ladies' Army and Navy Club was open to relatives of officers serving in the Army or the Navy. It attracted 3,500 members in its first year. It was still active in the mid-1920s.

Ladies' Associations for the Care of Friendless Girls

These Associations, established in 1870, each autonomous bodies with their own set of rules to suit the neighbourhood, were inspired by the work of their founder, Miss Ellice Hopkins, the moral and legal reformer. Their common aim was to raise the quality of life particularly for poor girls. Many activities were undertaken, including classes for mothers and teachers in training the young, forming girls' clubs in manufacturing towns, following up poor and friendless girls after they left school and placing them in respectable service and distributing suitable publications to teachers, parents and Christian workers. The Associations were also involved in encouraging emigration. Workhouse children were helped to make a good start in life and children rescued from unsatisfactory surroundings were sent to suitable schools. Members took a minute interest in every aspect of girls' lives, from missing a last train to criminal acts and regarded themselves as the girls' friend and counsellor. A yearly conference was arranged by the Preventive and Rescue Committee of the **National Union of Women Workers**. By the last decade of the nineteenth century, there were over 120 Associations in centres of population.

(Hopkins, 1878)

Ladies' Association of the Universities' Mission to Central Africa

The Mission's main task was to raise funds for sending out and supporting ladies, nurses and teachers for hospital and school work. In 1876 it was able to appoint a superintendent in Mozambique to oversee the Mission's activities.

Ladies' Athenaeum

A short-lived club, founded by Mrs Jennie Cornwallis-West (Lady Randolph Churchill) in 1913 at 31 Dover Street, London. It was intended for ladies interested in art, politics, literature and music. It no longer existed in 1928.

Ladies' Automobile Club

The Automobile Club of Great Britain and Ireland, later the Royal Automobile Club, moved to Brook Street, Piccadilly in 1903. It refused to admit ladies on account of the large numbers who wished to use the limited accommodation and 'the desire to permit smoking throughout the building'. A number of prominent ladies, headed by Lady Cecil Scott Montagu, were responsible for setting up the Ladies' Automobile Club at Hans Crescent Hotel, moving to Claridge's in the following year. The Club had six main objectives: to provide a social meeting place for motoring enthusiasts; to assist in acquiring technical knowledge; to provide driving instruction 'for members and their servants'; to organise tours and motors competitions; to provide accommodation for members' cars in 'motor houses'; and to secure foreign driving certificates and other requirements for touring abroad.

Ladies' Chess Club

Situated at 30a Sackville Street, Piccadilly, the Club provided facilities for practice on three days a week from 3 p.m. to 10.30 p.m. The entrance fee was 10s 6d and town membership £1 11s 6d. The Club arranged matches for the season and club competitions were arranged. In 1911, the San Remo International Tournament for ladies took place.

Ladies' Committee for the Amelioration of the Condition of Women in Heathen Countries

Founded in 1858 by Mrs Batchelor, who was working with young women in India, following a request for help. It later changed its name to the Wesleyan Women's Auxiliary. Some time elapsed before other female Methodist missionary societies were set up: the Bible Christian Women's Missionary League (1892), the Ladies' Missionary Association of the United Methodist Free Church (1897), the Methodist New Connexion Ladies' Missionary Association (1899) and the Primitive Methodist Women's Federation (1909). They were unified in 1930 and called 'Women's Work'. It co-ordinated its activities with those of the Methodist Missionary Society.

(Barrow, 1981)

Ladies' Concert Society

This is mentioned in the periodical *Shafts* in 1895; the organiser appears to have been Mrs Winifred Robinson, whose string quartet took part in four concerts arranged by the Society in November and December, 1895.

Ladies' Debating Club

This Club, which held its meetings in various venues, flourished in London in the last quarter of the nineteenth century. In 1877 it first admitted gentlemen for one evening per annum to participate in debates. In that year, the debate, held at St George's Hall, Langham Place, London was on the payment of Members of Parliament.

Ladies' Debating Society, Birmingham

The Society was founded in 1879 to conduct debates on matters of public interest. Early motions included: 'That education in elementary schools should be absolutely free'; 'That women should be allowed to enter the learned professions'; and 'That the influence of music on character is greater than that of painting'. Its first president was Mrs Ashford, who was succeeded by Mrs Brieley. In 1886 there were 55 members, mainly living in Edgbaston. It was still active in 1887.

Ladies' Dramatic Club

The object of this Club, situated at 126 Oxford Street, London, W, was to give an opportunity for talented amateurs to display their talents in the cause of charity. The secretary was Miss Nora Hastings, who ran it in connection with her Dramatic Academy and School of Elocution.

Ladies' Dressmaking Association

The Association had its headquarters at 42 Somerset Street, Portman Square, London and flourished in the last quarter of the nineteenth century. One of its specialities was the making of historical and national costumes for people who submitted their own designs and materials.

Ladies' Early Shopping Association

This Association was set up in 1885 after a series of drawing rooms in Kensington and Maida Vale. Its main purpose was to secure better working conditions for shop workers. The Association called on ladies not to shop later than 6 p.m. and asked shop proprietors to make strong efforts to establish a five-day week for its workforce. It suggested that there should be one early closing day and one late opening evening during the week. The Association also urged ladies to allow their servants and other members of their household to shop at an early hour.

(Hallsworth and Davies, 1910; Paine, 1912)

Ladies' Edinburgh Debating Society

Originally called the Edinburgh Essay Society, this organisation was founded in 1865 by Miss (later Dame) Sarah Elizabeth Siddons Mair, granddaughter of the actress Sarah Siddons, for young ladies eager for self-improvement. The Society met on the first Saturday of each month at her house, 29 Abercromby Place, though it later moved to 5 Chester Street, Edinburgh. The group started its own magazine, the *Attempt*, on topics such as croquet, dress fashion and Sunday opening of museums. In 1876, it was renamed the *Ladies' Edinburgh Magazine* and was sold commercially. The last issue to be published was in 1880. The Society had already changed its name to the Ladies' Edinburgh Literary Society and it was now once more changed to the Ladies' Edinburgh Debating Society. From its inception, debates were a regular feature. As early as 1866 the Society debated the merits of women having the parliamentary vote (of the 20 women only five were then in favour). Miss Mair was an ardent but non-militant supporter of women's franchise and the topic frequently appeared on the Society's agenda. Other subjects included: 'Do ghosts have an objective existence?', 'Should women be strong-minded?', boarding school versus home education, the opening of the professions, universities and the Civil Service to women and important social issues of the day. A number of members of the Society became well-known in their own right in later years, such as Charlotte Carmichael, later Mrs

Carmichael Stopes, mother of Marie and a distinguished Shakespearian scholar. Visitors to the Society included the Hon. Mrs Lionel Tollemache and Charlotte Yonge. Miss Mair was president of the Society from its beginnings, when she was 19, up to the closing of the Society in 1935 when she was 90.

(Rae, 1936)

Ladies' Educational Association

From 1869, this Association offered special classes for ladies at University College, London taught at separate times from students but by the same professors. They were also admitted to ordinary classes in two or three of the University's departments, including logic, political economy and the fine arts. The Association from its inception had many problems and it dissolved in 1878, when University College admitted women to degree courses.

Ladies' Empire Club

Founded in 1902 and situated at 69 Grosvenor Street, London, the Club, opened by Joseph Chamberlain, then Colonial Secretary, was the brainchild of the **Victoria League**, which promoted the travelling of students in all parts of the Empire. It also catered for ladies visiting London. In 1906 there were 700 members resident in the United Kingdom and 350 from the colonies. The Club was still in existence in 1950.

Ladies' Field Club

Situated at 32 Dover Street, London, W, this Club was for 'ladies interested in all kinds of country sports, pursuits and pastimes'. Election was by committee and the subscription was six guineas annually. In 1904, the secretary was Miss Geraldine Tronson.

Ladies' Golfing Union

By 1867, ladies were playing golf in Scotland and St Andrews Ladies' Golf Club became the first women's golfing club in the world. In the following year another club, at Westward Ho! was founded: here, no member of the club was allowed to bring a dog with her in case it should hurt or disturb the sheep grazing nearby. The growth in ladies'golfing clubs was spectacular: there were five clubs in 1878, 19 in 1884, 50 in 1894 and 220 by 1898. The latter figure was partly the result of the official recognition of golf as a woman's sport by the founding of the Ladies' Golfing Union, following a meeting at the Grand Hotel, Trafalgar Square, London in April 1893. The purpose of the Union was to promote the interests of the game, to obtain a uniformity of rules of the game by establishing a representative legislative authority, to establish a uniform system of handicapping, to act as a tribunal and court of reference on points of uncertainty, and to arrange the annual Championship Competition and obtain necessary funding for that purpose. The financial structure was based on the entrance fees and subscriptions from affiliated clubs. Much of this work was carried out by the dynamic first honorary secretary, Isette Person Miller, from her home in Putney. By the time of the Union's second annual general meeting, there were over 2,000 members. A *Golfer's Handbook* was compiled giving the rules of the game as well as a house journal called *Ladies' Golf.* By 1927 there were over 1,000 affiliated clubs and 400 Colonial clubs. Shortly after the LGU was set up, the Irish Ladies' Golf Union was formed after a meeting in Belfast. Scotland followed next in 1902 and Wales in 1904. As early as 1906 the formation of an English Ladies' Golf Union was considered, but this did not take place until 1957.

(Cossey, 1984)

Ladies Guild of Change Ringers

Founded in 1912, with an inaugural meeting at St Giles-in-the-Fields, London, it caters for both tower and handbell ringers. Its first honorary secretary and treasurer, from 1912–47, was Miss E. K. Parker (who became Mrs E. K. Fletcher), of 97 Edenbridge Road, Bush Hill Park, Enfield, Middlesex. Its initial object was to encourage ringing among Church workers as a physical and spiritual exercise, to assist when and where necessary their brethren in the art of ringing and to bring into closer relationship all women and girl ringers in the kingdom. In 1912 subscriptions were one shilling per annum for ringers, with a life subscription of five shillings for honorary non-ringing members. It was still active at the time of writing.

(Stevenson, 1999)

Ladies' Imperial Club

Situated at 17–18 Dover Street, London, this Club was for members of the Conservative or Unionist Parties. Residential accommodation was provided. The club flourished at least until the 1920s.

Ladies' Institute

Started by Bessie Rayner Parkes and Barbara Bodichon at Princes Street, Cavendish Square, London, it moved in 1860 to 19 Langham Place. The Institute was the offices of the *Waverley Journal* and subsequently of the *English Woman's Journal*, and these and other feminist magazines were made available to middle-class ladies. The club proved to be a useful

place for members to meet as it was open 11 hours a day and meals were also served. The subscription was one guinea a year, though governesses paid only half this amount. The Institute closed in 1867 but its aims and traditions were carried on in the **Berners** and **Somerville** Clubs.
(Crawford, 1999)

Ladies' Lacrosse Association

Lacrosse was first introduced into England from Canada in 1867 and by the end of the century was popular in girls' rather than in boys' schools. A Southern Ladies' Lacrosse Club was started in London in 1905; it consisted of former students of Wycombe Abbey, Roedean and Prior's Field Schools. The captain of the Club, Audrey Beaton, an old Roedeanian, subsequently established a Ladies' Lacrosse Association in 1912, mainly for schools, colleges and polytechnics. Beaton became honorary secretary and Penelope Lawrence, headmistress of Roedean, was president. In the following July, the Association issued administrative procedures and rules for international, national and county lacrosse matches. Membership of the Association was strictly limited to amateurs. By the end of the first season, there had been an increase from 20 to 70 schools and colleges as well as seven clubs. In 1913 the first international match was staged at Richmond cricket ground when England beat Wales and Scotland before an audience of 100 mainly women spectators. By the outbreak of the First World War the number of schools and colleges playing lacrosse rather than hockey had risen to 100 though it did not rival the latter game in popularity. Dress fashions were more innovative than in hockey. In 1901 the Ladies' Lacrosse Association encouraged players to wear tunics of 'regulation length – one inch off the ground when kneeling'.
(McCrone, 1988)

Ladies' Land League

Begun in early January 1881 at 39 Upper Sackville Street, Dublin by Anna Parnell, a younger sister of Charles Stuart Parnell, the Irish politician. Shortly before that, Anna with her sister, Fanny, had been involved in the United States of America in raising funds on behalf of the Land League. Subsequently, she was invited by the Land League Executive to form a Ladies' Land League (LLL) in Ireland. Parnell himself was amongst a minority of the Executive who strongly opposed the idea. The Executive had envisaged the LLL as a charitable, fund-raising movement but Anna and her colleagues saw it as a much more active and militant body. It would support evicted tenants, encourage resistance to land sequestration, erect wooden huts for the evicted and provide financial support for prisoners' families. The central body of the LLL consisted of a remarkable collection of women, including Mrs Anne Deane, the president and John Dillon's cousin, Hannah Reynolds, Mary O'Connor, Mrs A. M. Sullivan and the secretaries, Anne Stritch, Nan Lynch and Harriet Byrne. A 21-strong reserve 'Executive' was in place to take over in the event of arrests being made.

The LLL undertook an ambitious programme of activities. A detailed dossier of every Irish estate was compiled, the illicit printing of *United Ireland* was carried out, it co-ordinated activities such as ensuring a strong presence at evictions, and speeches, encouraging defiance to the authorities, were given to well-attended rural audiences. Disillusioned with the strategy adopted by the Land League, Anna accused members of timidity and hypocrisy. Branches of the LLL were quickly formed in Ireland and in London. Their activities led to several members, including Reynolds and O'Connor, being arrested, but many others took their place. The Land League was proscribed in October 1881 and W. E. Forster, the Chief Secretary, attempted a similar move with the LLL but did not carry out the threat. Anna had opposed the Land Bill of the same year and she came into conflict with the views of the Land League over the tactics to be employed. The LLL became more extreme. Determined to carry on the Land War, Anna made tours of Scotland and north-east England denouncing Gladstone's attitude towards Ireland. She received much notorious publicity in June 1882 when she flung herself before the horse of Lord Spencer, the Lord-Lieutenant, on his way through Dublin and, holding the bridle, harangued him on the plight of tenants evicted near Limerick. Relations between the Land League and the LLL deteriorated as the extent of the latter's financial extravagance became clear. In July 1882, their overdraft reached the sum of £5,000. Anna was shattered by the death of her sister Fanny that month and Parnell was furious at the LLL's extremist attitude; in the following month, the Ladies' Land League was dissolved with little resistance being offered.

(Bew, 1979; Foster, 1976)

Ladies' National Association for the Repeal of the Contagious Diseases Acts

Formed in 1869 under the leadership of Josephine Butler to oppose the state regulation of prostitution, in response to calls for the extension of the Contagious Diseases Acts to the whole of the United Kingdom, this was probably the most conspicuous organisation campaigning for the repeal of the Acts, in which it was finally successful in 1886. As its name implies, its membership was restricted to women; this may have been partly because when the National Association for the Repeal of the Contagious

Diseases Acts was founded a few weeks earlier, there had been no women present, probably due to an oversight. The weekly organ of both bodies was the *Shield*, which was published from 1870 to 1886. It organised meetings and published a wide variety of campaign materials both independently and in conjunction with other bodies campaigning for the repeal of the Acts. It was also involved in a number of by-elections in which it propagandised against candidates known to support the Acts; on these occasions its members were sometimes subjected to threats of violence. After the repeal of the Acts in 1886 the Association continued campaigning against regulated prostitution, changing its name to the Ladies' National Association for the Abolition of State Regulation of Vice, or simply the Ladies' National Association. From the 1890s on it was particularly concerned with regulation of prostitution in the Empire, especially in India under the provisions of the Cantonment Acts and the Contagious Diseases Act (India), but still involved itself with issues of women's sexual subjection in Britain. In 1915 it amalgamated with the British Branch of the **British, Continental and General Federation for the Abolition of Government Regulation of Prostitution** to form the **Association for Moral and Social Hygiene**.

(McHugh, 1980; Walkowitz, 1980)

Ladies' National Silk Association

This Association was formed for the purpose of enlisting the co-operation of ladies 'in restoring the benefits of this ancient industry to the silk operatives of the British Isles, and in reviving a means of support for so many of their countrymen and countrywomen'. By 1895, the Association had over 3,000 members, many of them influential ladies, and had as its president the Duchess of Teck. Functioning on a county basis with honorary county secretaries, the Association encouraged its members to promote British silk manufacturing by purchasing native-made goods in preference to others.

Ladies' Negro Friend Society

Founded in 1825, the Society, which had a strong Christian element, circulated petitions, collected money and diffused literature, drawing attention nationally to the plight of slaves, and campaigned for their freedom. The Birmingham Society was the leader in this movement, and other Societies were formed in Great Britain. The abolition of slavery in 1838 was greeted as a triumph by the Society though it continued its work until at least the last two decades of the nineteenth century. William Wilberforce, the great reformer, expressed his dislike of the Society. 'I own,' he wrote, 'I cannot relish the plan. All private exertions for such an object becomes their character, but for ladies to meet, to publish, to go from house to house stirring up petitions, these appear to me proceedings unsuited to the female character as delineated in Scripture. And though we should limit the interference of our ladies to the cause of justice and humanity, I fear its tendency would be to mix them in all the multiform warfare of political life.' The cause was closely connected with the lack of rights of women at the time, especially in the United States of America. In 1836, an Englishwoman, Angelina Grimké, was invited to New York to hold private meetings with Christian women and later addressed an audience of women. The following year, she was again invited to America and with her sister attracted large numbers at their meetings in Massachusetts. Later, the Society was active in other parts of the world, corresponding with friends of abolition in Zanzibar, with the Scottish missionaries at Lake Nyasa and it appealed to the government to insist on the emancipation of slaves in Cuba. It also worked closely with the British and Foreign Anti-Slavery Society. A Hibernian branch had been established in Ireland in 1830.

Ladies' Rifle Club

This Club flourished at the beginning of the century with its headquarters at Brighton. Bisley rules and regulation targets were used. The entry fee in 1912 was 10 shillings. Clubs sprang up around the country and in 1910 the National Rifle Association admitted the first Ladies' Club to affiliate. A Ladies' Bisley was subsequently held with members using the Service rifle and competing on equal terms with men.

Ladies' Sanitary Association

Established in 1857 by Miss Rose Adams in London, to popularise sanitary knowledge among the poor by means of simple tracts on health, hygiene and domestic subjects, by lectures and the provision of libraries. Courses were also set up for schoolmistresses on practical and theoretical aspects of health education. Branch associations were opened throughout the country and attracted middle-class ladies who wished to help in the work. The Association over the next few years expanded its activities to include the distribution of cleaning materials and the establishment of sewing and cookery classes; visits were made to houses where families could be assisted. By 1887, over 1,400,000 leaflets had been produced and distributed. The development of the Health Visiting Service in 1891 and the introduction in the same year of women inspectors under the Factory and Workshops Act signalled the end of amateur philanthropy in this field. It survived until 1900, and in 1908 the Health Visitors' Service became a municipal undertaking.

(Dowling, 1963)

Ladies' Sanitary Reform Association of Manchester and Salford

The Association, founded in 1862 to promote principles of health and cleanliness, was staffed initially by voluntary workers. It was soon divided into districts, each having a lady superintendent, a paid health visitor and support from voluntary workers. By 1905 there were 23 districts. No assistance was given by the sanitary authorities until 1890, when the Manchester staff were placed under central direction and the salaries of some visitors were paid by the sanitary authorities. One interesting feature of the Association was that it anticipated by 45 years the establishment of a 'School for Mothers'; these were early welfare clinics run by voluntary associations (the first official one was at St Pancras in 1905) where weekly meetings of mothers were organised. Health talks and practical demonstrations were given on such topics as sick nursing, the making of baby clothes and cooking. The work of the Association stimulated interest in other towns and a number of local authorities subsequently appointed women sanitary inspectors whose work included health visiting.

Ladies' Scottish Climbing Club

Founded in 1911, this Club was open to membership for those who had scaled four peaks of over 3,000 feet, preferably in Scotland (two climbs in snow were not included in the above) and two rock climbs of sufficient difficulty. Candidates also had to have done some climbing during the previous three years. The honorary secretary was Miss Inglis Clark. During 1909, one of its members, Mrs L. S. Gibbs, scaled Mount Kinabalu, the highest peak in the Malay Archipelago, rising to 13,455 feet, its first ascent by a woman.

Ladies' Work Society

For ladies who were in need, the Society provided them with a livelihood by arranging to send their needlework and embroidered objects for sale to the Society's London headquarters in Sloane Street. The Society also endeavoured to raise the standard of needlework. Royal patronage was provided by its president, Princess Louise, Marchioness of Lorne, who took a great personal interest in the work produced.

Lady Almoners

In 1895, Mary Stewart, the secretary of the North Pancras District Offices of the Charitable Organisation Society and a trained social worker, was appointed to the Royal Free Hospital, London. Her task was to prevent abuse of the use of the hospital's outpatients department by those able to pay for treatment elsewhere. After initial opposition to the appointment by some of the medical staff, the experiment proved a success and the scheme was extended to ward patients in 1897. When Westminster Hospital adopted the system two years later, the almoner, Helen Nussey, established contacts with various agencies outside the hospital who could assist patients, such as the clergy, charities and doctors in private practice. By 1902, seven of the London hospitals had made almoner appointments. In 1903, Helen Mudd, almoner at St George's Hospital, was responsible for forming a Council of Almoners to discuss common difficulties and problems. No officers were elected, but monthly meetings were held. The need for training was recognised early and, from 1907, members were awarded diplomas on courses, some of which were social-science based.

Provincial cities had now appointed almoners and, in 1912, all certificated almoners were invited to join the Almoners' Council. In 1922, the Council, an offshoot of the Charity Organisation Society, was reconstituted as the Institute of Hospital Almoners; two-thirds of its Executive Committee members were from outside the profession. Two years earlier, an Association of Hospital Almoners, consisting of trained almoners and independent of the COS, was formed. The two bodies existed side by side until 1946. The Poor Law Act of 1928, which removed the functions of the Poor Law from the Local Government Board and gave powers to local authorities to take control of infirmaries, increased the work of almoners. Some almoners were now receiving training in mental welfare work and some almoners split off and joined the Association of Psychiatric Social Workers. In 1946, with the National Health Service reorganisation on the horizon, the two professional bodies, the Institute and the Association of Hospital Almoners, merged to become the Institute of Almoners, as a self-governing profession. It changed its name in 1964 to the Institute of Medical Social Workers.

(Bell, 1961; Lochhead, 1949)

Lady Cyclists' Association

This Association, with its headquarters at 35 Victoria Street, London, SW, was set up to provide rides, tours and other social gatherings for lady cyclists. It was affiliated to the National Cyclists' Union, which catered for both men and women and which had been established in 1878. It published a monthly journal, the *Lady Cyclists' Association News*, and a handbook (priced 4d.) which contained reliable information about country inns, where ladies could be made sure of being comfortable on moderate terms. Some of the members wore rational dress (see **Rational Dress Society**). The entrance fee was one shilling and the subscription was 3s 6d annually. In 1900, the president was the Countess of Malmesbury and the honorary secretary was Miss Grace Murrell.

Lady Golfers' Club

Following the founding of the **Ladies' Golfing Union** in 1893, attempts were made to establish a club in London. Negotiations were entered into with the **Victoria Club** so that LGU members could join either as full members or temporarily at reduced rates, but this initiative did not prove possible. In 1911, with the growth of affiliated clubs and the need for more office space, the LGU formed a Lady Golfers' Club at 3 Whitehall Court. The Club contained a large room for meetings, luxurious furnishings and sleeping accommodation. On at least one occasion, when the weather was bad, a putting competition took place on a nine-hole course laid out in a large empty room. After the First World War, the dining room was described as 'a rest cure for those who had eaten to jazz music'. In the 1960s, it merged with the Golfers' Club.

(Cossey, 1984)

Lady Guides' Association

Formed by Miss Edith A. Davis in 1889 to provide occupation for qualified ladies with a practical knowledge of history and features of places of interest for visitors to London and other parts of the country. The guides arranged railway tours, seats for concerts and lectures, met travellers at railway stations, taking them to their hotels and acting as couriers. All candidates had to sit an examination before being allowed to work. Its headquarters was at 16 Cockspur Street, Charing Cross, London.

Lancashire and Cheshire Women Textile and Other Workers' Representation Committee

It was formed in 1903 by Sarah Reddish, Selina Cooper, Sarah Dickenson, Esther Roper and Eva Gore-Booth, with the original purpose of sponsoring an election candidate who favoured women's suffrage. The candidate eventually chosen who fought the 1905 election in Wigan for Labour was Thorley Smith; he was beaten by the Conservative candidate. In 1906, the Committee sent representatives to London to take part in a joint demonstration with the **Manchester and Salford Women's Trade and Labour Council** and the **Central Society for Women's Suffrage**. In the first 1910 election, the Committee combined with the **National Industrial and Professional Women's Suffrage Society** to sponsor a women's suffrage candidate, again unsuccessfully. It was still in existence in 1913.

(Liddington and Norris, 1978)

Land Army *see* **Women's Land Army**

League of Isis

This organisation was possibly initiated by Frances Swiney in about 1907, although in the second edition of her *The Bar of Isis* (1909) she attributes its origin to Henry Ancketill of Durban, Natal. Its object was to promote the protection of motherhood and more rational and humane sex relations generally; it had strong theosophical leanings. It met regularly at the Eustace Miles Restaurant, London, and by 1911 had branches in New York, South Africa and India. From 1912 it gained publicity in the *Eye-Opener* and the *Awakener*. It does not appear to have survived the First World War.

League of Jewish Women

Established in 1943 during the Second World War, the League concerns itself with social service and social action both within the Jewish community and in the wider community. Voluntary work undertaken includes hospital and residential home visiting and assisting the aged and sick in their own homes. Day centres cater for a range of activities and a Friendship Club movement was started in 1949. At school level, the League liaises with pupils with special needs and with speech clinics, pre-school playgroups and nurseries. At the time of its 50th anniversary in 1993, the League had 71 groups and more than 5,000 members. It publishes twice yearly its journal *Around the League*.

League of Justice

It was in existence in 1913 to carry out a 'programme of uncompromising militancy without violence or law-breaking to win the vote and break up unjust monopolies'.

(Crawford, 1999)

League of Skilled Housecraft

Established in 1922, the League aimed at making domestic service more attractive to women and girls at a time when there was a severe shortage of domestics. Since its inception, the **Girls' Friendly Society** had been concerned with the welfare and moral education of domestic servants; it called a special conference in December 1921, which concluded that 'the question of making domestic work a profession, with training qualifications and status [was] in the interests of the country'. A committee, consisting of members of a number of organisations, was subsequently set up and it produced a scheme which would meet this need.

The League of Skilled Housecraft began its work with 100 girls taking the preliminary examination at the end of 1922. Graduates of the

League could wear a badge and uniform though few chose to do so. Within three years, the League was experiencing organisational problems. By 1925, the **Girls' Friendly Society** was solely responsible for the organisation though a few years later a reformed organisation proved to be more successful. However, by 1942, the League was conducting its courses in evening classes which were open to girls aged 16 and over, and there were no examinations. The unpopularity of domestic service, especially in wartime when many alternative types of work were available to women, proved fatal to the League's success and it ceased to exist in 1972.

(Horn, 2000)

League of the Church Militant

This was formed from the **Church League for Women's Suffrage** in 1917. Its emphasis rapidly shifted from the vote to women's ordination. Its most prominent member was Maude Royden. It appears to have ceased activity in 1928. Its organ was the *Church Militant*.

Liberal Women's Suffrage Union

It was formed in 1913 as a successor to the **Forward Suffrage Union** to express Liberal suffragists' dissatisfaction with Liberal policies, and as a reaction to the new pro-Labour policies of the **National Union of Women's Suffrage Societies**. Its most prominent member seems to have been Eleanor Acland, wife of the MP Francis Acland. It appears to have ceased activity after 1917.

(Crawford, 1999)

Ling Association

The first School Board for London decided shortly after its establishment in 1870 to promote physical education in its elementary schools by means of military drill. Miss Martina Bergman (later Bergman-Österberg) was appointed as Superintendent Teacher of Physical Education in 1881. She had been trained in Swedish gymnastics, a system consisting of a progressive series of ordered exercises relating to the physical growth of the child, which had been formulated by Henrik Ling, founder of the Stockholm Gymnastics Institute. By the late nineteenth century, the system formed part of girls' high-schools curriculum in many parts of Britain.

Bergman-Österberg founded a training college in Hampstead in 1895 which moved to Dartford and was given the title the Bergman-Österberg Training College, the first residential college for women physical-education teachers. Her philosophy was encapsulated in a speech: 'Let us once and for all discard men as physical trainers of women ... This work we would do better, as our very success in training depends upon our having felt like women, able to calculate the possibilities of our sex, knowing our weaknesses and our strength.' In January 1899, 31 of her former students met in Hampstead to form a Ling Association. Membership was open to those who had trained with Bergman-Österberg or who had trained at the Stockholm Institute. Its purpose was to provide a forum for promoting the system and to compile a register of those qualified to teach it. By 1904, the Association was conducting its own examinations and offering a Diploma in Physical Training. Before the First World War it was also concerned with the salary and status of physical-education teachers. By the 1930s, the system was considered to be too restrictive, with the need for more emphasis rather to be placed on recreation, keeping fit and later, dance. Nevertheless, the Ling Association laid the foundation for the expansion of gymnastics in the twentieth century.

(McIntosh, 1968)

Local Government Advancement Association

Under the chairmanship of Mrs Humphry Ward, this Association set out to educate municipal women voters to take a more active interest in the affairs of their own town or district. It encouraged members to stand against anti-suffrage candidates for municipal bodies and to demonstrate that women could influence parliamentary decisions without necessarily having the vote. The Association was active during the first two decades of the twentieth century.

London and Country Club

This Club, at 81 and 82 Holland Park, London, W, was for 'ladies of good social position', chiefly those engaged in literary and professional work. Although a residential Club, a limited number of non-residents were also elected. The annual subscription was two guineas.

London and National Society for Women's Service

After the passing of the Enfranchisement of the People Act in 1918, the **London Society for Women's Suffrage** decided to shift its focus to a wider range of feminist issues, particularly around employment, and in 1919 changed its name to the London and National Society for Women's Service. In this it was following the lead set up by its own Women's Service Bureau during the War, the secretary of which, Philippa Strachey, became the first secretary of the newly redefined Society. This, with premises in Marsham Street, Westminster, which became Women's Service

House, rapidly established itself as a vigorous organisation campaigning particularly for professional women. Although originally an affiliate of the **National Union of Societies for Equal Citizenship**, it broke with it in the mid-1920s over a number of issues, particularly family endowment. Its activity in finding women's professional employment was eventually delegated to the **Women's Employment Federation**. A major feature of the Society's activities was the events that took place at Women's Service House, especially those organised by the Society's Junior Council from 1929 to 1940. This included a programme of talks by a variety of speakers, most notably including Ethel Smyth and Virginia Woolf. Additionally, from 1926 the Society's Library (later the Fawcett Library) was developed by the Librarian, Vera Douie, into a significant research resource.

On the outbreak of the Second World War, the Library was evacuated to the Home Students' College (now St Anne's College), Oxford. In 1940 Women's Service House suffered bomb damage, and the Society had to make do with temporary offices in nearby Tufton Street (from 1949 the Library was based in the Great Smith Street branch of Westminster City Libraries). In 1953 the Society changed its name to the Fawcett Society in honour of Millicent Garrett Fawcett. In 1957 it finally moved to new premises in Wilfred Street, Westminster. In the 1960s and 1970s it was active in campaigning for women's equality, and was particularly involved in the Sex Discrimination Act of 1975, whose mover, Baroness Seear, was president of the Society. In the early 1970s it also sponsored a women's liberation periodical, *Women's Report*. However, in 1977, financial pressures compelled it to leave its Wilfred Street premises, and resulted in the separation of the Library from the Society. The Library was transferred to the then City of London Polytechnic, now London Guildhall University, while the Society has continued as a campaigning organisation for women's equality.

(Strachey, 1927; *Women's Studies International Forum*, 1987)

London and Westminster Tailoresses' Society

Established in 1879 by Emma Paterson, an early champion of women's trade unionism. She became its first secretary. The Society consisted of 134 women working for the Royal Army Clothing Depot and had its head-quarters in Bloomsbury. It became a branch of the **London Tailoresses' Trade Union** in 1883 and was represented at the Trades Union Congress. Its membership remained low throughout its existence (by 1900 there were only 17 members) and the Society was dissolved in 1903.

London Biblewomen and Nurses' Mission

This Mission, originally called the Bible and Female Domestic Mission, and established in 1857 by Mrs Ellen Henrietta Ranyard, was one of the pioneering organisations for women's work among the poor in London. After observing the 'misery of our sisters there' in the parish of St Giles in London, she began a society for training Biblewomen as mission workers who were engaged in house-to-house visitations. They were given six months' instruction in Bible study, hygiene, charity organisation and the different principles of relief giving. The women worked under the direct superintendence of the clergy. In 1868, a Nursing Branch was formed, but was quite distinct from the Biblewomen. The candidates received a two years' hospital training and were sent, in some cases, to maternity hospitals, then going on to the hostel of the Mission at 47 Parker Street, Kingsway, London, WC1 for training in district work. Those who successfully completed the course became Ranyard Nurses. Subsequently their work was monitored by experienced Superintending Sisters. The Missions endeavoured to encourage those who were earnest Christians to apply for the nursing course. Both Biblewomen and nurses were recruited from women between the ages of 25 and 35. After the death of Mrs Ranyard in 1879, the name of the organisation was changed to the London Biblewomen and Nurses' Mission. The movement rapidly spread to other parts of the country.

(*Official Year-Book of the Church of England,* 1906; Ranyard, 1875)

London Council for Welfare of Women and Girls and Central Information Bureau

The object of the Bureau, which was founded in 1922, was to bring together the various voluntary agencies concerned with the welfare of women in the metropolitan area, to monitor the provision in London for women in moral danger and to ensure their spiritual, mental and physical welfare. A Central Information Bureau for Girls which gave advice on accommodation, social clubs, classes which could be attended and other matters was set up and its services were used by women, parents and social workers both in London and in the rest of the country.

London Girls' Club

Founded by Flora Lucy Freeman towards the end of the nineteenth century for working-class girls. An annual bazaar was held in which the girls participated.

(Freeman, 1901)

London Graduates' Union for Women's Suffrage

Established in 1909 by Louisa Garrett Anderson, it organised graduates of London University, and those closely associated with the University. It appears to have ceased activity after 1918.

(A. J. R., 1913)

London Ladies' Shorthand Association

With its headquarters at Craven Road, Lancaster Gate, London, this Association, started in 1889, was formed to meet the demand for public men and journalists who employed lady secretaries. Applicants seeking employment through the Association needed to be able to write 100 words per minute using Pitman's shorthand. For a weekly engagement with hours of 10 a.m. to 4 p.m., the pay was 35 shillings, and for morning sessions, 15 shillings.

London National Society for Women's Suffrage

In 1866, the **Enfranchisement of Women Committee** was formed in London with the object of bringing together supporters of women's suffrage, and it collaborated with the Manchester committee in organising the Women's Suffrage Petition of that year. In 1867, the committee was dissolved, and the London National Society was formed under the leadership of Mrs P. A. (Clementia) Taylor. This joined with the equally newly formed **Edinburgh National Society** and **Manchester National Society** to form the **National Society for Women's Suffrage**. However, in 1872, the London Society disagreed with the other major societies over support for the campaign against the Contagious Diseases Acts, and declined to co-operate with the new co-ordinating Central Committee of the National Society for Women's Suffrage until 1877. In the 1888 split of the National Society for Women's Suffrage, the London Society was also split between the two Central Societies. It only re-emerged in 1900 as the new **Central Society for Women's Suffrage**.

(Rover, 1967)

London Provisional Petition Committee

Originally named the Women's Suffrage Petition Committee, this Committee was formed in 1866 by women of the Langham Place Circle. Its work was soon taken over by the **Enfranchisement of Women Committee**.

(Crawford, 1999)

London Pupil Teachers' Association: Girls' Division

Founded in 1887 with Millicent Garrett Fawcett as its first president, the Association's aim was to 'give pupil teachers interests and pleasures apart from their school life, and to enable them to gain a higher conception of the privileges and responsibilities of their calling'. This was done in a number of ways. 'At homes' were held for small groups by 'women of culture', reading parties were formed as well as glee clubs and game clubs, and visits were paid to universities, galleries and museums. Its headquarters were at Toynbee Hall and Henrietta Barnett succeeded Mrs Fawcett as president in 1891. By 1892, there were 1,220 pupil teacher members attending activities in 12 centres in different parts of London. Many distinguished educationists helped with the work.

London Society for Women's Suffrage

The successor to the former **Central Society for Women's Suffrage**, although with a more restricted area of operation and no national co-ordinating function, it adopted the name of London Society in 1907. Its secretary was Philippa Strachey; Lady Frances Balfour was president, and Millicent Fawcett was chairman till 1909. The Society was based in Victoria Street, London, SW. It was probably the most prominent of the constituent societies of the **National Union of Women's Suffrage Societies**, often acting nationally on its own initiative. Its independence was most clearly marked during the First World War, when it immediately set up the Women's Service Bureau (under the auspices of the National Union of Women's Suffrage Societies) and sponsored the initiative of the **Scottish Women's Hospitals** under Elsie Inglis to send medical assistance to France, Serbia, Romania and Russia. In 1919 it became the **London and National Society for Women's Service**.

(Rover, 1967)

London Tailoresses' Trade Union

No date is known for its foundation but the **London and Westminster Tailoresses' Society** became a branch of the Union in 1883. One of its aims was to join the Amalgamated Society of Tailors, who, by 1895, took an interest in organising women tailors. It is probable that it was finally absorbed into the Amalgamated Society.

Lyceum Club

Founded in 1903 by the writer Constance Smedley because of dissatisfaction with the **Writers' Club** as a venue for meeting and networking, it was open to women of any nationality who had either published any original work in literature, journalism, science, the arts or music, or had university qualifications or, initially, were wives or daughters of distinguished men. Its object was to focus the work of women in the arts, science, medicine, music, public service, journalism, drama and other important directions, as well as to afford members a congenial space in which to meet editors, agents and the like. Social members, who were in strong and active sympathy with the aims of the Club but were not qualified professionally, were admitted on special terms. It was not until 1904 that the Club moved into its first premises, formerly those of the defunct Imperial Service Club, at 128 Piccadilly; its first chairman was Lady Frances Balfour, who was succeeded, some time prior to 1920, by Lady Aberdeen; her vice-chairman was Enid Moberly Bell, and the first honorary secretary was Constance Smedley. Special features of the Club at least before 1920 included an Art Gallery, where exhibitions of highest standing of members' work was on view; monthly concerts, when original compositions were performed by distinguished artists; and a book gallery, where members' books were sold. There was also a Bureau which 'renders services in many ways to professional members'. Initially there were 35 bedrooms.

Members paid an entrance fee on joining, and an annual subscription. In 1909 the rates were: town, four guineas, entrance fee, two guineas; country, three guineas, entrance fee, one guinea. Rates for social members were higher. Its members were arranged in 'sections': agriculturalists and horticulturists; authors; arts and crafts; dramatists; geographical; journalists; music; painters; photographers; public service; social; and university. Most activities were arranged in 'circles', which would be addressed by 'lecturers of repute and experience'. By 1920 the circles comprised: American, Belgium [*sic*], French, Irish, Italian, Oriental, Polish, Russian, Scottish, Irish, United Empire, Art Lovers, Geographical, Music, Photographers, Poetry, Billiard [*sic*], Bridge and Psychic. A major general activity was the Club's Monday House Dinners, with eminent after-dinner speakers, at which 'the most celebrated representatives of the arts, and the Ambassadors and Embassies of various countries', were entertained. There were fortnightly dances on alternate Friday evenings 8.30–11.30 p.m., tickets 5s. Debates were held weekly. Some sample motions from 1920: 'That the Turk should be compelled to leave Europe'; 'That it is desirable that England, as well as America, should go dry' (the proposition was resoundingly defeated); 'That autobiographical self-revelation is justifiable'. The Club's organs were the *Lyceum Journal* and the *Lyceum Annual*.

In 1919 the Club's original premises were sold to the RAF Club, and the Club moved to 138 Piccadilly, apparently a very imposing house: 'its decorations include a *chef d'oeuvre* of the Adam Brothers and paintings reputed to be by Angelica Kaufmann in the ceilings and panels respectively of the Dining Room'. However, the Club's finances seem not to have been equal to its ambitions, and by 1934 it had moved to 9 Chesterfield Gardens, W1. It was still active in November 1934.

Associated clubs included a Lyceum Alpine Club, founded in 1907, 'the only Alpine Club in the world exclusively for women', and international daughter clubs: there were Lyceum Clubs in Amsterdam, Athens, Barbados, Basel, Berlin, Berne, Brussels, Florence, Geneva, Hobart, Lausanne, Madrid, Melbourne, Milan, Nice, Paris, Rome, Stockholm and Sydney.

(Smedley, 1929)

Male Electors' League for Women's Suffrage

It was formed probably in 1896 by Ben Elmy, husband of Elizabeth Wolstenholme Elmy. It was probably the first specifically male organisation in support of women's suffrage; in 1907 one of its members, Bayard Simms, was the first man to be imprisoned for militant women's suffrage activity.

(Crawford, 1999)

Manchester and Salford Society of Women in the Bookbinding and Printing Trades

Founded in 1896 by Isabel Forsyth, a printing worker, it campaigned principally for improved pay for women in the binding and printing trades, its initial goal being a wage of £1 per week, which it achieved in 1917. By 1924 it had nearly 2,500 members. It later became known as the Society of Women. In 1942 it amalgamated with the National Union of Printing, Bookbinding and Paper Workers.

(Tidcombe, 1996)

Manchester and Salford Women's Trade and Labour Council

Its origin was in a 1904 split in the Manchester and Salford Women's Trade Union Council, which had been provoked by Christabel Pankhurst over the issue of women's suffrage. As well as campaigning for women's suffrage, it took up such traditional trade-union activities as forming a women's insurance society. The secretaries were Sarah Dickenson and Eva Gore-Booth. It appears to have ceased functioning in 1914.

(Liddington and Norris, 1978)

Manchester Ladies' Literary Society

Lydia Becker, the prominent campaigner for women's suffrage, who was born in Manchester, became president of this Society in 1867. It presented the opportunity for women to meet regularly and be prepared to discuss important matters of the day. In her presidential address, Becker, conscious of the lack of opportunity of women to obtain higher education, outlined the advantages of a trained mind. As author of a book *Botany for Beginners*, published three years earlier, she urged a study of physics and science as a good basis for intellectual nourishment.

(Spender, 1983)

Manchester National Society for Women's Suffrage

In 1865 (according to Elizabeth Wolstenholme Elmy) a Manchester Committee for Women's Suffrage was formed, and in 1867 this became the Manchester National Society for Women's Suffrage. It joined with the Edinburgh and London Societies to form the **National Society for Women's Suffrage**. As the Manchester Society for Women's Suffrage, it affiliated to the **National Union of Women's Suffrage Societies** in 1897 and continued until 1918.

(Crawford, 1999; Pankhurst, 1931)

Manchester Society of Women Painters

Few opportunities existed for the training of women artists in the nineteenth century. This Society was formed in 1879 'to provide facilities for the members working together and studying from the life' and 'to disseminate the principles of true art among art students generally'. Art classes in connection with the Society included elementary drawing, drawing from the antique and drawing and painting from a 'living model'. It also held exhibitions of works for sale by members of the Society. One member had exhibited at the Royal Academy and another at the Salon in Paris.

Married Women's Association

It was formed in 1938 on an initiative by Edith Summerskill, who became its first president (subsequent presidents have included Vera Brittain and Juanita Frances; other prominent members originally included Helena Normanton and Lady Helen Nutting). Its first offices were at 20 Buckingham Street, London, WC2. Its original objects were to promote legislation to regulate the financial relations between husband and wife as between equal partners; to secure for the mother and children a legal right to a share in the marital home; to secure equal guardianship rights for both parents; and to extend the National Insurance Acts to include women on the same terms as men. Later objects were to extend family allowances, to establish equal pay and to awaken women to their full political responsibilities. In 1952 a major disagreement, nominally over Helena Normanton's evidence to the Royal Commission on Marriage and Divorce, led to the departure of a number of prominent members, and the establishment of the **Council of Married Women**. The Association's organs have been *Wife and Citizen* (1945–51), and in more recent years the *MWA Newsletter*. The Association is still active at the time of writing.

Married Women's Property Committee

It was formed in April 1868, originally with the object of agitating for the passage of the Married Women's Property Bill, and more generally to abolish the legal disabilities married women experienced with regard to their property. It was based in Manchester, and its joint secretaries were Josephine Butler and Elizabeth Wolstenholme (Elmy). Prominent members of the executive committee included Lydia Becker, Jacob Bright, his wife Ursula Bright, Dr Richard Pankhurst, Frances Power Cobbe, Lilias Ashworth, Lady Amberley, Peter Taylor and his wife Clementia ('Mentia'), Sir Charles Dilke, Isabella Tod, Emilie Venturi, and, from 1879, the young Mrs Emmeline Pankhurst. It was wound up on 8 November 1882, subsequent to the final passage of the Married Women's Property Act.

(Holcombe, 1983)

Matrons' Council of Great Britain and Ireland

Established by Ethel Bedford Fenwick, a former matron at St Bartholomew's Hospital, London, with Isla Stewart and five other matrons in July 1894, the Council consisted of matrons from England, Scotland, Ireland and Wales. It acted as a forum for discussing matters of common concern. These included a scheme to form an International Council of Nurses and the need to reform the army nursing service, leading to the creation of the **Queen Alexandra's Royal Army Nursing Corps**. The Council held annual conferences from 1898 and became affiliated to the **National Union of Women Workers**. In 1904, it co-operated with the **Society for the State Registration of Trained Nurses** and other organisations to form the **National Council of Nurses of the United Kingdom**. By 1926, it had only 126 members.

Medical Women's Federation

In 1877, the Royal Free Hospital, London, had admitted women medical students for clinical training and by 1880 there were 20 women on the Medical Register. In the same year, with Dr Elizabeth Blackwell as president, an Association of Registered Medical Women in London was formed as well as one for the provinces. In 1916 the different associations came together to form the Medical Women's Federation, to improve the conditions of employment of women doctors. It was incorporated as a limited company in 1917 and by the following year it had 385 members. In the interwar period, the Federation adopted many important causes, such as the need for lunacy reform, and the level of maternity mortality and venereal disease. It championed instruction in birth control for women and advocated the reshaping of the gynaecological curriculum. It holds annual conferences and mounts refresher courses. By 1980, there were 3,226 members. More recent issues which the Federation have raised have

concerned sex education in schools, abortion reform and the effect of the National Health Service reorganisation on women doctors. It issues a quarterly journal *Medical Woman*.
(Bell, 1953)

Men's Committee for Justice to Women

It was founded in 1909 to watch cases in which women were prosecuted in suffrage connections. Prominent members were Albert Dawson and Captain C. Gonne.
(Crawford, 1999)

Men's League for Opposing Woman Suffrage

Formed early in 1910, it was merged later in that year in the **National League for Opposing Woman Suffrage**.
(Harrison, 1978)

Men's League for Women's Suffrage

Established in 1907 by Herbert Jacobs, with the Earl of Lytton as president, the League engaged principally in propagandist work. In theory it supported all suffragists, but in practice leaned heavily towards the militants. Prominent members included the journalists Henry Woodd Nevinson and Henry Brailsford, as well as the novelist and playwright John Galsworthy. It had its own organ, the *Men's League for Women's Suffrage Monthly Paper*, which ran during 1909–14. The League appears to have ceased activity in 1914, although individual members continued campaigning until 1918.
(John and Eustance, 1997)

Men's Political Union for Women's Enfranchisement

Founded in 1910, this was, in intention at least, the male affiliate of the **Women's Social and Political Union**. Its members frequently engaged in militant action, and were imprisoned. Prominent members included Victor Duval and his brother-in-law Hugh Franklin, Henry Harben and Henry Woodd Nevinson. Its activity ceased on the outbreak of the First World War in August 1914.
(John and Eustance, 1997)

Men's Society for Women's Rights

Formed by A. W. Gray Jamrach in 1912, this was largely concerned with the traffic in women, or the white slave trade. It also supported militant women's suffrage action, especially that of the **Women's Social and Political Union**. Prominent members included Captain C. Gonne,

William Boulting and Frank Witty. Until June 1914 it was closely associated with the weeklies the *Eye-Opener* and the *Awakener*, edited by Jamrach until his links with the WSPU were deemed too close. It seems not to have been active after 1916.

(Crawford, 1999)

Methodist Women's Association

Formed in Belfast in 1884, this Association consists of a women's committee which seeks fellowship and links with other churches in Northern Ireland and the Republic and with other organisations to co-operate in attaining peace and reconciliation in Ireland. It mounts education programmes and carries out charity work and is responsible for 50 branches of the **Young Women's Christian Association** in Ireland. The Association has approximately 5,000 members.

Metropolitan and National Nursing Association

William Rathbone, a Liverpool philanthropist and politician, had financed a scheme of district nursing in his city which was devoted to looking after the sick in their own homes. In 1862, he founded the Liverpool Training School for Nurses. He was also one of the founders of the National Association, begun in 1874, 'for providing Trained Nurses for the Sick Poor in London and Elsewhere' under a Lady Superintendent, Mrs Dacre Craven. Its aim was to train and supply district nurses on a national scale. It changed its name to the Metropolitan and National Nursing Association in the following year with its training school in Bloomsbury Square, London. It became affiliated to the Queen Victoria's Jubilee Institute for Nurses, later the **Queen's Nursing Institute,** in 1887 though it retained its own Council.

(Owen, 1965; Stocks, 1960)

Metropolitan Association for Befriending Young Servants

Established in 1875, following a report by Mrs Jane Nassau Senior, daughter-in-law of the economist and the first women Inspector of Workhouses and Workhouse District Pauper Schools, on the education of girls in pauper schools. Mrs Senior drew attention to the insufficient preparation of the friendless workhouse girls for adult life and recommended that lady visitors, banded in an Association, should supervise them on behalf of the Guardians from 13 years of age until they reached 20. The Association built 22 training and lodging homes and provided suitable training for service. In 1893, there were 32 district committees and some 1,100 ladies were involved in the work. By 1897, over 8,000 girls had been found places. Homeless girls were accommodated whilst awaiting situations. The Association, which operated only in London, worked in conjunction with the **Girls' Friendly Society**, who offered similar help in the provinces.

(Pratt, 1898)

Metropolitan Board Mistresses' Association

Established in 1882 to resist a proposal brought to the School Board for London to exclude married women from teaching. The Association appears to have been disbanded thereafter but was revived in 1897 as the London Mistresses' Association after the Board produced a scheme to raise the salaries of men but not women teachers. Ultimately, the Board withdrew the proposal and increased women teachers' salaries.

(*Governess*, 1882)

Minerva Club

Begun in 1920 by Dr Elizabeth Knight, a champion of extending the franchise for women, at 28 Brunswick Square, London. It was a Club for members of the **Women's Freedom League**, a militant suffrage organisation, and was open to both men and women. The Club promoted vegetarianism in its restaurant. The Club was also a meeting place for the Women's Freedom League Executive. Elizabeth Knight died in 1933 but the Club remained in existence until 1962.

(Crawford, 1999)

Mothers' Union

The Union was begun by Mrs Mary Sumner in 1876, following a meeting of mothers in the rectory at Old Elmsford. By 1892, the Union was represented in 28 dioceses and had a membership of 62,000. The movement rapidly spread abroad and branches were opened in many of the Colonies. In 1913, the Union was incorporated as a Church Society with three objects: to uphold the sanctity of marriage; to awaken in all mothers a sense of their great responsibility in the training of boys and girls; and to organise in every place a band of brothers to unite in prayer. The Union produced two quarterly periodicals, the *Mothers' Union Journal* and *Mothers in Council*, edited by C. M. Yonge, for upper-class members. The constitution was completely revised in 1972 when the purpose of the Mothers' Union was stated to be 'specifically concerned with all that strengthens and preserves marriage and the Christian family life'. Women who declared their support for the objects of Union and had been baptised were eligible, thus allowing single and divorced people to join.

National Advisory Centre on Careers for Women *see* Women's Employment Federation

National Assembly of Women

This was initiated by a march in London on 9 March 1952, ostensibly for women's rights in general, but mainly campaigning against British involvement in the Korean War. Other concerns in 1952 included the importation of American comics and increased nursery charges and school dinner prices. Later concerns included colonialism and women in developing countries. It was noticeably sympathetic to the Communist Party and the Soviet Union. It appears to have ceased activity in 1981.

National Association for Maternal and Child Welfare

The Association, formed in 1911, was concerned with improving the health and well-being of mothers and children by liaising with health authorities, schools and voluntary organisations. It offered health education programmes, health conferences and it conducted research.

National Association for Promoting the Medical Education of Women

In order to provide a medical education for women in Edinburgh, a committee was formed in 1871 which became the National Association for Promoting the Medical Education of Women. Its campaigning was successful and led to women being admitted to medical schools. In the years 1887–88, the number of registered medical women rose from 54 to 72 with increasing numbers preparing for the profession. In the same year, the degree of Doctor of Medicine of the University of London was granted for the first time to a women, Dr Mary Scharlieb. The Association, having achieved its main aims, was wound up in 1889.

(Jex-Blake, 1887)

National Association for Supplying Female Medical Aid to the Women of India

Six months after her arrival in India in January 1885, the Marchioness of Dufferin and Ava, the wife of the new Viceroy, at the Queen's request announced the formation of this Association. Its main objective was to provide medical tuition, including organising Indian women as doctors, nurses and midwives, providing medical aid and dispensaries and cottage hospitals for the treatment of women and children, and to supply female medical officers and nurses for women's hospitals. The Association made grants in aid to Indian medical schools. In 1891, there were 224 European and native women under medical instruction; 48 hospitals were connected with the fund and 466,000 women had attended them. On her return to England in 1889, Lady Dufferin organised the United Kingdom branch of the Association which assisted the Association in India, provided Indian scholarships in British schools of medicine and made grants available to students preparing for work in India.

National Association of Ladies' Circles of Great Britain and Ireland

Established in 1936 to support the Round Table movement, the Circles are open to women under 45 years of age. They support the Round Table by fund raising, and by involvement in community service projects. There are local meetings and social activities throughout the year. Their organ is the *Circle* and there are approximately 5,000 members.

National Association of Training Corps for Girls

This Association was formed during the Second World War in 1942 after national service for women was introduced. Its specific aim was the training of young women between the ages of 14 to 20 to help with the war effort, either by preparing them to enter one of the Services, hospitals, munition factories or to work on the land. The programme consisted of a basic training for all cadets, which included first aid, fire-fighting, hygiene and drill; specialised training of their choice, such as aircraft recognition and car maintenance; service to the community, such as canteen duties and knitting for the Services; and charity work, usually for those connected with the war. Training was rigorous and all units met twice a week. Uniforms were always worn. The Association attracted many thousands of girls.

(Cooke, 1946)

National Association of Women Civil Servants

Established in 1932 from an amalgamation of the **Federation of Women Civil Servants** and the Civil Service section of the **Association of Women Clerks and Secretaries**. Besides campaigning for equal pay, the Association attempted to obtain recognition for women who were not members of the general grade organisations. Its direct rival, the Civil Service Clerical Association, after attempting to bring about an amalgamation in 1936, set out to weaken the National Association's influence. By 1958, its influence was at a low ebb, and when it ended in 1961 there were a little over 2,000 members.

National Association of Women Pharmacists

By 1913, there were about 200 qualified women chemists, of whom almost two-thirds were employed as dispensers in public institutions and doctors' surgeries, with approximately 20 in business on their own account. The great majority belonged to the Association of Women

Pharmacists which was founded in 1904. The objects of the Association were to maintain a register of all qualified women pharmacists and of the posts which they held, to discuss ethical questions relating to their work at regular meetings, and to act as a means of finding employment for members and to examine conditions of employment. The Association introduced a special scheme to encourage insurance for old age. Its name is now the National Association of Women Pharmacists.

National Association of Women's Clubs

The Association was set up in 1935 to provide facilities for bringing women together socially and assist in organising informal education courses. Individuals clubs met once a week and their programmes ranged from cookery demonstrations and homemaking to discussions on aspects of central and local government and international affairs. The Association encourages community and charity work and promotes self-development through education and recreation. It maintains day schools and workshops for members. Groupings of individual clubs into country or city associations were made for the purpose of planning conferences, but each club was autonomous. An annual conference was held, and contact with government departments and other organisations was maintained on issues affecting women.

National Association of Women's Lodging Homes

Its object was to link together all organisations interested in improving and providing lodging accommodation for working women. Members visited lodging houses, distributing tracts to women inmates and urging them to attended church services. Its honorary general treasurer was Mrs Hylton Dale, 60 Onslow Gardens, London, SW. It was active in about 1910–12.

National Baby Week Council

The Council was established in 1917 by Miss Halford, who organised the first Baby Week to highlight the need for better child-care and welfare and made a case for a Ministry of Health. Because of its success, the Council became a permanent body and it undertook a wide range of activities. These consisted of Baby Weeks, Baby Days, film displays, lectures, posters and publications. In 1933 it organised no fewer than 244 Baby Shows and Exhibitions, and held competitions for the best baby clothes, and essays and lullabies, both in words and music. It offered an annual Challenge Shield for the most effective piece of propaganda in the United Kingdom and also an Imperial Baby Week Challenge Shield for good work done in the Dominions and Colonies.

(McCleary, 1935)

National Birth Control Association

This was formed as the National Birth Control Council in April 1930 with the original object of co-ordinating the activities of the five existing birth control societies: the **Society for Constructive Birth Control and Racial Progress**, the **Birth Control Investigation Committee**, the Birth Control International Information Centre and the **Workers' Birth Control Group** (in November 1933 the first of these withdrew, but by 1938 all the rest had fully merged with the Association). Although it had branches nationwide, its head office was in London. It changed its name in July 1931 to allow new clinics and branches to join without representation on the governing body. Its hierarchy was extremely eminent, with Lady Denman as chairman, Sir Thomas Horder as president, an array of extremely distinguished vice-presidents, and an Executive Committee whose members included Eva Hubback, Margery Spring Rice, Mary Stocks, Marie Stopes and Helena Wright. The secretary was Margaret Pyke. Drs Norman Haire, Carlos Paton Blacker and Helena Wright were among the expert advisers.

Its principal stated aims were to enable especially poor married women to avoid unwanted pregnancies, but many of its members had wider aspirations, and saw birth control as a means of giving women in general a degree of control over their own fertility. It rapidly developed from being merely a co-ordinating body into providing its own clinics and campaigning for birth-control provision in general. Despite such major fundraising events as gala performances, the Association's finances remained in a poor way, and it regularly had to be bailed out by wealthy members (mainly Lady Denman). During the 1930s it was careful to avoid any association with campaigns to decriminalise abortion, in 1936 going so far as to forbid one of its branches affiliating to the newly-formed **Abortion Law Reform Association**. Equally, it distanced itself from campaigns for the sterilisation of the mentally unfit. In 1939 the Association revised its objects and changed its name to the **Family Planning Association**.

(Leathard, 1980)

National Board of Catholic Women

Established in 1938, its object was to act as the leading voice of Catholic women in England and Wales, and to stimulate discussion and action. The Board consists of representatives from many Catholic organisations having women members and there is a quarterly general assembly. As a consultative body to the Bishops Conference of England and Wales, it advises on matters of importance to Catholic women.

National British Women's Total Abstinence Union

Formed out of the reconciliation between the Women's Total Abstinence Union and the **British Women's Temperance Association** in 1925, the Union provides activities for children from the age of seven and young people up to the age of 30 which highlight the danger of alcohol. Weekend and summer schools are arranged as well as examinations and competitions. The *White Ribbon* was its journal. The Union is also concerned with the spread of gambling and the need for moral welfare.

National Childbirth Trust

The Trust was established in 1956 and prepares couples for childbirth and parenting. Ante- and post-natal classes are held, together with post-natal support groups. The Trust also conducts educational lectures for schools, colleges and other groups. The Trust's magazine, *New Generation*, is issued quarterly.

National Council for Equal Citizenship

This was formed in 1932 to continue the lobbying activity of the **National Union of Societies for Equal Citizenship**. In 1947 it amalgamated with the **National Women Citizens' Association**.

National Council for the Unmarried Mother and her Child

In 1918 at the end of the First World War, the National Council was formed to act as a co-ordinating body for the voluntary and statutory agencies concerned with the relief, protection and welfare of unmarried mothers and their children. It was active in a number of related fields: it kept a watch on legislation affecting the illegitimate child and put forward suggestions for appropriate new legislation; encouraged the provision of homes and hostels for the mother and her child; promoted public understanding of the problems of illegitimacy; and assisted social workers in obtaining grants or legal advice for the unmarried mothers they were helping. It later became the National Council for One Parent Families.

National Council of Nurses of the United Kingdom

In May 1902, the **Matrons' Council of Great Britain and Ireland**, which supported the notion of state registration, relinquished the conduct of the campaign for state registration to a new body, the Society for the Registration of Trained Nurses. By the time of its first meeting, there were over 400 members. Its president was Louisa Stevenson, a member of the Board of Management of the Royal Infirmary, Edinburgh. The Society was responsible for the introduction of a bill in the Commons in 1904 which would have allowed state registration. In the same year, it merged with the **Matrons' Council of Great Britain and Ireland** and other organisations to become the National Council of Nurses of the United Kingdom.

The Council was a self-governing body and its object was to promote mutual understanding and unity between the various associations of nurses in Great Britain. Any association of qualified nurses with a membership of not less than 50 was eligible for affiliation with the Council. It looked to improve training and to promote the interests of its members. By its affiliation with the International Council of Nurses, the Council gained a knowledge of nursing conditions in other countries. Its headquarters was at 17 Holland Place, London, W.

(Baly, 1995; Pavey, 1951)

National Council of Women *see* National Union of Women Workers

National Federation of Business and Professional Women's Clubs of Great Britain and Northern Ireland

It was founded in 1938 by Beatrice Gordon Holmes and Lena Madesin Phillips, in close association with Caroline Haslett and Phyllis Deakin. From the outset it was linked with the International Federation of Business and Professional Women, whose representative Zonola Longstreth was also involved in its beginnings. The Federation started with three London-based clubs: Westminster, City of London and Thames, which were soon joined by another London club, the Regency, as well as clubs in Bedford, Sheffield, Halifax, Newcastle, Leeds, Nottingham, Chesterfield and Manchester. It did not acquire headquarters premises until June 1940, at Stratton House, Stratton Street, London, which soon suffered bomb damage so that the office had to be moved to 6 Suffolk Street, London, SW1, and in 1942 to 71 Fleet Street, London, WC1. The Federation aimed at encouraging business and professional women to realise their responsibilities in their own cities, their country and in world affairs. It hoped to promote effective co-operation between business and professional women throughout the world, to improve their status, raise their standards of education and training and to work for the removal of sex discrimination. Members were urged to stand as local councillors, magistrates and Members of Parliament and to influence management and policy-making in business by obtaining suitable posts. Its organ, begun after the war, was the *Business and Professional Woman*. In the 1980s it changed its name to the United Kingdom Federation of Business and Professional Women's Clubs.

(Hall, 1963)

National Federation of Townswomen's Guilds *see* **Townswomen's Guilds**

National Federation of Women Teachers

Following the rejection by the National Union of Teachers in 1904 of a call for equal pay, an Equal Pay League was formed in order to secure the election of women to the Union's Executive. Its name was changed in 1906 to the National Federation of Women Teachers and it soon adopted the suffrage cause. Three years later the London unit of the NFWT together with another organisation within the NUT, the **Women Teachers' Franchise Union**, broke away from the main Union to form the **National Union of Women Teachers**.

(Gosden, 1972; Tropp, 1957)

National Federation of Women Workers

Despite the fact that the **Women's Trade Union League** had done much to organise female labour since its inception in 1874, some 30 years later it was clear that many trades were still without any representation and the average earnings of women workers compared unfavourably with men. Mary Macarthur, then the secretary of the League, established the National Federation of Women Workers in 1906 to remedy the situation. The Federation was organised as a general labour union with a central body and local branches. Membership was restricted to women from unorganised trades and those who had not been admitted to their appropriate trade union. The League played a substantial part in the Federation's running, providing three members to the executive committee. Mary Macarthur was its first president, becoming general secretary in 1908. She was also editor of the League's journal, the *Woman Worker*.

The Federation was a recognised trade union and was affiliated to the Trades Union Congress and the General Federation of Trade Unions. By the end of 1906 it had recruited over 2,000 members in 17 branches in England and Scotland. The London group of women's societies was absorbed in the new organisation. Industrial unrest in 1910, when women from a variety of industries joined men strikers, resulted in a great boost to membership. Soon afterwards Susan Lawrence, then a member of the London County Council, urged charwomen employed in London elementary schools to form a branch of the Federation; this resulted in a considerable bettering of their conditions of pay and service. Lawrence then joined the Federation and became one of its most distinguished members.

Like the **Women's Trade Union League**, the Federation supported the principle of co-operation with the skilled men's societies, especially those working in the same trades. Although this resulted in a more influential organisation, the Federation was weakened by many branches transferring to men's societies, as in the case of the Society of Bleachers, Dyers and Finishers. In other instances, branches survived but with considerably smaller membership. However, one measure of the Federation's success in organising women labour was the degree of hostility shown by employers towards it. By 1916 the Federation had a membership of 20,000 in many miscellaneous trades including confectionery, ammunition, chain-making and laundry work. In 1920, the Federation merged with the General Workers Union.

(Boston, 1987)

National Federation of Women's Institutes

These rural women's organisations were originally a Canadian invention; the first was established in 1897 at Stoney Creek, British Columbia. Despite many efforts to introduce them, the idea of rural women's organisations was resisted in Britain until 1915, when agriculture was being organised on a war footing. The main initiative in Britain was taken by Mrs Alfred Watts who, after many official rebuffs, eventually found an interested partner in the Agricultural Organisation Society, a body with government subsidy. The first name chosen for the new bodies was to have been Women's Co-operative Institutes, but this was dropped because of likely confusion with the existing **Women's Co-operative Guild**.

In 1915 three Institutes were set up in Wales, and one in England; the movement rapidly developed, and in 1916 many more local institutes were started, so that by 1917 it was felt necessary to form a National Federation of Women's Institutes. The first chairman of this body was Lady Denman, who continued to be effectively the national face of the Women's Institutes until her death in 1954. The Institutes were not always popular with the women of the local squirearchy, because of their insistence on democratic procedures ('one member, one vote'). Even so, they traditionally had difficulties in preserving the balance between their democratic ideals and their more hierarchical instincts; it was not until 1961 that they had a non-titled chairman.

Their original aims were to stimulate interest in agriculture; to develop local co-operative industries as well as home and local industries in general; to study home economics; and to provide centres for educational and social intercourse and for local activities. In practice, it is the last of these aims with which the Women's Institute movement has become identified. They very quickly became a natural feature of the rural scene, to the extent that they are taken for granted as part of village life in the

1920s in such works as E. M. Delafield's *Diary of a Provincial Lady* (originally published in weekly parts in *Time and Tide*). The cliché associates the Institutes with 'Jam and Jerusalem': the jam connection is mainly due to a government initiative in the Second World War as a means of keeping the women busy (and distracted from taking too much interest in more serious matters), but the devotion to Sir Hubert Parry's setting of the Blake verse usually known as *Jerusalem* has a long radical feminist history (it was claimed in 1918 by Millicent Fawcett as 'the women voters' hymn'). Interestingly, in recent decades, the women's movement has tended to ignore the Women's Institutes, which is rather strange: with a national membership of more than 250,000, the National Federation of Women's Institutes is far and away the largest women's organisation in Britain. No other women's organisation needs to hire the Albert Hall for its annual general meeting. Also, in recent years the Institutes have been careful to demonstrate their concern for political, and particularly feminist, issues, although not always in a way that the orthodox women's movement finds acceptable – for example, discussions over the regulation of prostitution. Despite the continuing decrease in agricultural employment, the Women's Institute maintains its place as Britain's most representative women's organisation.

(Andrews, 1997)

National Free Church Women's Council

The Council promoted the ideals of the United Free Church by holding services but also aimed at co-operating with other Christian committees in serving the community. Founded in 1908, the Council operated through local Councils throughout the country, carrying out such functions as hospital visiting, establishing homes and flatlets for the elderly and hostels for single mothers.

National Health Society

This Society, founded in 1873 by Dr Elizabeth Blackwell, was formed with the object of diffusing sanitary and hygienic knowledge. To achieve this, simple lectures were given in working men's clubs and mothers' meetings on such topics as air, ventilation, food and cookery, hygienic dress, smoke abatement, the provision of open spaces and the prevention of the spread of diseases. Teaching of the laws of health were encouraged in schools by offering prizes to both teachers and pupils. The Society's motto 'Prevention is better than cure' well expressed its objective. After 1891, special courses of training lectures were organised by the Society for ladies wishing to qualify as teachers of hygiene and nursing under the County Council schemes of Technical Education, or as factory or sanitary inspectors. By 1895, a staff of over 30 ladies gave lectures under the auspices of County Councils and also published and distributed many leaflets and pamphlets. At this stage, it could boast of three Princesses amongst its patronesses.

National Housewives' Register

The Register was founded in 1960 on an initiative of Mary Nicol with the aim of offering housebound wives 'with liberal interests and a desire to remain individuals' the opportunity to meet in each others' homes in order to break down isolation. A large range of topics was discussed and regional and national conferences were arranged which offered a chance of wider exchanges. By 1981 it had 1,200 groups comprising approximately 25,000 members. A biannual newsletter was also published. In 1987 it changed its name to the National Women's Register, reflecting its broader appeal to all women.

(Jerman, 1981)

National Industrial and Professional Women's Suffrage Society

The Society was the national body representing the Lancashire and Cheshire Women Textile and Other Workers' Representation Committee and the Manchester and Salford Women's Trade and Labour Council; its secretary was Esther Roper, and Eva Gore-Booth was equally prominent.

(Lewis, 1988; Liddington and Norris, 1978)

National Institute for Promoting the Employment of Women in the Art of Printing

This Institute was based in Edinburgh, and from 1862–65 appears to have published the magazine *The Rose, the Shamrock and the Thistle: a journal for the fair daughters of Great Britain and Ireland.*

National Institute of House Workers

During the Second World War, a governmental committee, chaired by Violet Markham, was set up to examine the implications of the severe shortage of domestic help, particularly where professional women were carrying a 'double burden'. Shortly after the war, Markham established the National Institute of House Workers to promote domestic service as a skilled craft.

(Lewis, 1984)

National Joint Committee of Working Women's Organisations

This organisation, situated at the Labour Party headquarters, was set up in 1916 'to represent the women of the political, industrial and co-operative movements of the workers.' It aims to secure their representation on local, national and international bodies which deal with matters of special interest to women. The Committee, consisting of 50 members, organises meetings and conferences, seeks publicity and campaigns to achieve this end. It is required to maintain effective liaison with women members of the Parliamentary Labour Party. The Committee meets four times or more a year, and it requires the approval of affiliated organisations before action can be taken. Originally called the Standing Joint Committee of Industrial Women's Organisations, the word 'Industrial' in the title was replaced by 'Working' in 1941. The present name dates from 1952.

National Labour Women's League

Although women had been active supporters of the Labour movement from the 1880s, it was not until a meeting at the home of J. Ramsay MacDonald, then secretary of the National Labour Representative Committee, and his wife Margaret in March 1906 that action was taken. An organisation of Labour women was formed, 'to work for Labour Representation in connection with the Labour Party'. At its first Conference three months later, the League's objectives were formulated. Members would support candidates at local and parliamentary elections, work for full rights of citizenship for all women and men, educate themselves by means of meetings and discussions, become involved in the work of Poor Law Guardians and educational institutions and stand as candidates for town and district councillors and MPs. Margaret MacDonald was its president and Mary Macpherson its secretary. Initially there was no direct representation of the League within the Party, but in 1909 Margaret Bondfield became the first League representative at a Labour Party Conference.

The League campaigned on the question of sweated industries and higher education for women and men by means of special leaflets which they wrote and distributed and through its journal *Labour Woman*. It was also active in the international field, sending delegates to the Socialist and Trade Women's Conference in Stuttgart in 1907 out of which emerged the International Women's Bureau. Before the First World War, a small number of the League's members had been elected to local government bodies. By 1913, there were five borough councillors and 19 members of Boards of Guardians. The total membership of the League was only 6,000 and it faced financial difficulties. From 1918 the situation changed and candidates were successful on a larger scale. The first Labour woman chairman of the London County Council, Alderman Mrs Lowe, was elected, but it was not until the 1923 general election that the first three Labour women became MPs, Susan Lawrence at East Ham

North, Margaret Bondfield in Northampton and Dorothy Jewson in Norwich.

During the First World War the expansion of women's employment made it necessary to draw together Labour and industrial organisations into a central consultative committee, the Standing Joint Committee of Industrial Women's Organisations (later the **National Joint Committee of Working Women's Organisations**). One of its main concerns was equal pay, as seen in the campaign in 1919 to achieve it for men and women teachers, and in the Civil Service in 1928. Other issues which were pursued included divorce-law reform, suffrage and the entitlement of women to join trades unions relevant to their employment. Five years later, the League issued a report highlighting the effect of poor nutrition on maternal health in areas of high unemployment. With the achievement of parliamentary enfranchisement of women from 1918, the Women's Labour League merged with the Labour Party. The activities of the League was given a new impetus in 1918 when the Party granted women members the right to create Women's Sections for special work among women electors. By 1932, no fewer than 2,000 such Sections had been set up in Britain. It became the Labour Party Women's Advisory Committee in 1951 and the National Labour Women's Committee in 1981. It consists of 26 women, both Labour Party National Executive nominees and representatives of the Party's women's organisations and is located at the Labour Party's headquarters.

(Middleton, 1977)

National League for Opposing Woman Suffrage

Formed in 1910, it brought together the Women's and Men's Anti-Suffrage Leagues, under the leadership first of the Earl of Cromer, then of Lord Curzon and Lord Weardale. It took over from the Women's League the *Anti-Suffrage Review*, which ceased in 1918, together with the organisation.

(Harrison, 1978)

National Organisation of Girls' Clubs

Formed in 1911 by an amalgamation of the Girls' Clubs and the Rest Rooms Committee of the **National Union of Women Workers** and the Clubs Industrial Association, founded by the Organisations Committee of the **Women's Industrial Council**. The National Organisation hoped to develop the social, industrial, educational and physical possibilities for girls in a number of ways: by informing them of the scope of existing agencies, supplying a list of volunteer workers to help with clubs, arranging conferences and lectures for club leaders and by forming small reading parties for club members in association with the Workers' Educational Association.

National Political League for Men and Women

Founded by Miss Mary A. Broadhurst in 1911, it aimed to support political reform in general and women's suffrage in particular. It arranged meetings and publicised women's suffrage in general. During the First World War it organised the National Land Council to train women to work on the land. In 1917 it became the National Political Reform League.

(A. J. R., 1913; Crawford, 1999)

National Society for Women's Suffrage

It was founded in 1867 as a loose amalgamation of the Edinburgh, London and Manchester Societies for Women's Suffrage under the leadership of Lydia Becker; the Birmingham and Bristol Societies affiliated in the following year. From 1870–90 its organ was the *Women's Suffrage Journal*. In 1872 a Central Committee was inaugurated to co-ordinate activity throughout the United Kingdom, but immediately faced problems over the question of support for the Contagious Diseases Acts, and the **London National Society for Women's Suffrage** remained apart from the work of the Central Committee until 1877. In 1888 there was another division in the ranks, arising from the affiliation of societies with objectives other than that of suffrage, most notably the **Women's National Liberal Association**; the dissidents formed the Central National Society for Women's Suffrage, often referred to as the Parliament Street Society, as distinct from the Central Committee, the Great College Street Society. Especially subsequent to the death of Lydia Becker, the suffrage societies lacked a central co-ordination until the formation of the **National Union of Women's Suffrage Societies** in 1897.

(Rover, 1967)

National Society of Women Organisers

Women organisers were employed by Conservative women's branches which were set up in constituencies after 1919. Two years later, the National Society of Conservative and Unionist Agents agreed to allow women to become agents but the constituency agents rejected this move. In April 1927, the National Society of Women Organisers was established. Its aim was not to take over agents' work but to encourage women to become efficient Women District Secretaries. However, a number passed the appropriate examinations and did the work of agents. The Society was suspended during the Second World War. After the War, there were 70 members, of whom 20 were agents. The Society merged with the National

Society of Conservative and Unionist Agents and women became equal members.

(Lovenduski, Norris and Burness, 1994)

National Spinsters' Pensions Association

Founded in 1935, its aims were to obtain pensions at 55 years of age for spinsters (i.e. unmarried women) who contributed to National Health Insurance, and to obtain a contributory pension scheme for spinsters who did not at the time come under the National Health Insurance Scheme; it considered that this was necessary to compensate for discrimination experienced by spinsters as compared with men, married women and widows. In 1936 it claimed over 30,000 members in 55 branches, with national headquarters at Bradford, Yorkshire, the residence of the honorary national organiser and secretary, Miss Florence White, who was the moving spirit behind the organisation; the national president was Lady Fisher-Smith, the chairman was Isabel Forsyth and the secretary to its Parliamentary Committee was Mr J. C. M. Guy, MP. Although its main period of activity was in the late 1930s, the Association was still active in the 1950s.

National Union of Railway Women's Guilds

The Union, which was in existence in the 1920s, was formed to afford means of social intercourse amongst the wives and daughters of railway workers; to render such assistance to any of its members as might be necessary; and to co-operate with local branches of the National Union of Railwaymen, and with other women workers.

National Union of Societies for Equal Citizenship

Although formed in 1919 as the successor to the **National Union of Women's Suffrage Societies**, this had its own rather different agenda. Although not neglecting the 1920s campaign for equal franchise, culminating in the Equal Franchise Act of 1928, its concerns went far beyond the vote, including such New Feminist issues as maternal welfare, family endowment and, at arm's length, birth-control information, as well as such traditional feminist concerns as women's employment. Although officially non-party, its political line was generally left-of-centre, tending to favour Labour, albeit with strong reservations. Its organ was the *Woman's Leader*, successor to the *Common Cause*, with Eva Hubback as editor. Its leader throughout the 1920s was Eleanor Rathbone; she was elected to Parliament in 1929, and in the changing circumstances of the following years in 1932 NUSEC decided to separate its campaigning and educational functions, the former going to the new **National Council for**

Equal Citizenship, the latter, with the *Woman's Leader*, being delegated to the **Townswomen's Guilds**.

(Beddoe, 1989; Harrison, 1987; Law, 1997; Pugh, 1992)

National Union of Women Teachers

Founded in 1909 in order to achieve parity of status with male teachers, the Union was formed by two women's pressure groups, the London section of the **National Federation of Women Teachers** and the **Women Teachers' Franchise Union**. Never numerically very large, the NUWT was mainly concerned with promoting feminist causes, particularly the granting of the franchise on the same terms as men, and the securing of equal pay. By 1920, there were 21,000 members. It formally broke away from the National Union of Teachers that year when it resolved that candidates for the NUWT's Central Council could not be members of the NUT. Its organ was the *Woman Teacher*. In 1927, the Union supported the **Federation of Women Civil Servants** in demanding a select committee to inquire into the cost and method of the principles of equal pay in the Civil Service. It was wound up after gaining equal pay in 1961. At this time, the president was Miss J. M. Collins, and its general secretary was Miss Muriel Pierotti.

(Copelman, 1996; Phipps, 1928)

National Union of Women Workers (1)

This was set up in Bristol in 1874, following Emma Paterson's address to the National Association for the Promotion of Social Science's Congress there, to organise women workers. It does not appear to have been long-lived. It should not be confused with the following entry.

(Lewenhak, 1977)

National Union of Women Workers (2)

This should not be confused with the previous entry. By the late 1880s, there were several organisations, consisting of women engaged in some form of philanthropic work, who were involved in campaigning for women's rights in relation to, amongst other causes, married women's property rights, employers' liability, criminal law and trades unions. A Ladies' Union of Workers among Women and Children was established 'to form a common centre for all Ladies and Associations engaged in the work of helping and caring for women and children, and to encourage sympathy and co-operation'. In 1888, the Aberdeen Ladies' Union organised a Women's Conference of Women's Work, attended by over 1,000 women. Other conferences followed at Barnsley (1889), Birmingham (1890) and Liverpool (1891). At this last conference, a Central Council of the Conferences of Women Workers was established with a governing council and an executive committee. At the 1895 meeting it was decided to adopt a new constitution, and the organisation was to be known as the National Union of Women Workers. In October of that year, its first Conference was held at Nottingham under the presidency of Mrs Louisa Creighton. The vice-presidents included Lady Frederick Cavendish and Lady Laura Ridding. Beatrice Webb was an early member of the Executive. Its objectives were very clearly set out: 'The encouragement of sympathy of thought and purpose among the women of Great Britain and Ireland; the promotion of their social, civil and religious welfare; the gathering and distribution of serviceable information; the federation of women's organisations and the formation of local Councils and Union of Workers.'

There was some confusion at first as an earlier organisation with the same title as the NUWW, established in 1874, had been essentially for working-class women and was mainly concerned with female unionisation and legislative protection for workers. The new NUWW, on the other hand, defined women workers as 'women who are engaged in some real work whether it be paid or voluntary and who do that work, not for the sake of their own satisfaction, not for the sake of pay, though some may need some pay, but because they look upon their work as a vocation'.

Annual conferences, lasting three or four days, were held in different towns and cities and reports of their proceedings were published. The range of topics discussed was wide, touching on such matters as education, temperance, emigration, paid work for women, the problems of poor families and the promotion of social reform. The suffrage question was frequently debated: the Union provided a useful forum for both suffragists, such as Millicent Fawcett, and anti-suffragists, like Mrs Humphry Ward, both of whom were members. The Union's headquarters was in Berners Street, London, and a magazine, *An Occasional Paper,* was published quarterly. The NUWW was a great success; by 1912 it had attracted more than 7,000 members, operating through local branches throughout the United Kingdom.

At the meeting in October 1918 it was decided to change its name to the National Council of Women of Great Britain and Ireland and ten years later to the National Council of Women of Great Britain. The Council acts as a pressure group, working at national and international levels, to highlight the many issues concerning women in modern society.

(Glick, 1995)

National Union of Women's Suffrage Societies

Formed in 1897 to heal the split in the suffrage movement over the affiliation of Women's Liberal Associations, it superseded the Central

National Society for Women's Suffrage and the Central Committee of the National Society for Women's Suffrage, and brought together under one umbrella the Central and Western Society for Women's Suffrage, the Central and East of England Society for Women's Suffrage, North of England Society for Women's Suffrage and other provincial societies. The original executive included such prominent and various feminists as Lady Frances Balfour and Eva Gore-Booth, but the uncontested leader of the NUWSS throughout its existence was Millicent Garrett Fawcett.

The NUWSS had from its beginning a relatively loose structure; local societies were members of the NUWSS, but had a considerable degree of autonomy. In 1907 a federal structure, which had been evolving over the years, was recognised by the effective demotion of the **Central Society for Women's Suffrage** from its national co-ordinating function, and the grouping of local societies within, eventually, 18 regional federations. By 1914, the NUWSS had over 400 constituent societies. From 1907–09 it had heavily subsidised the weekly *Women's Franchise*, but from April 1909 its own organ was the *Common Cause*.

The response of the NUWSS to the militant action of the **Women's Social and Political Union** and the **Women's Freedom League** was at first ambivalent; indeed, Millicent Fawcett attracted much criticism in 1906 for giving a celebratory banquet for the first group of militant prisoners in London on their release from Holloway Prison. However, as militancy escalated, the NUWSS increasingly distanced itself from it, and in 1912 it finally adopted officially the terms 'non-militant, law-abiding' as part of its platform, in addition to its 'non-party' status.

In terms of party politics, although originally leaning towards the Liberals, the NUWSS also attracted a considerable number of Conservative women, and increasingly women with Labour sympathies. By 1912, the leadership had lost faith in the Liberal government to the extent that it formed a limited electoral alliance with the Labour Party, in return for the Labour Party's unqualified manifesto support for women's suffrage. To this end it set up its Election Fighting Fund, under the direction of Catherine Marshall, to support pro-suffrage, effectively Labour, candidates standing in by-elections when the opposing candidates had not made a specific commitment to women's suffrage; predictably, this policy caused increasing dissatisfaction among members, many of them prominent, who lacked sympathy with the Labour Party, and it is uncertain how well it would have lasted had a general election been called before August 1914. By that date the NUWSS had over 40,000 members.

In late 1914 and in 1915, the NUWSS gave low priority to specific suffrage campaigning in favour of war-related activity. However, the leadership's line on the war caused a major split, and many prominent members, such as Helena Swanwick, Catherine Marshall and Maude Royden resigned from the Executive in 1915 over the issue of promoting women's participation in war work, and over the leadership's negative attitude to the 1915 meeting of the International Woman Suffrage Alliance in The Hague.

The work of the NUWSS culminated in the Representation of the People Act of 1918, and in 1919 it re-formed itself as the **National Union of Societies for Equal Citizenship**.

(Holton, 1986; Hume, 1982; Rubenstein, 1991)

National Women Citizens' Association

Local Women Citizens' Associations began to be formed from 1913 following the initiative of Eleanor Rathbone in the Liverpool and Manchester areas. These differed from the **Women's Local Government Society** in that they were essentially autonomous local organisations rather than national organisations with local branches, but also in that at this time the Women's Local Government Society was perceived to be unsupportive of the campaign for women's parliamentary vote. When, in 1917, it became apparent that women would be granted a limited parliamentary franchise, it was widely felt that there was a need to educate women for citizenship and to stimulate their interest in it; to this end the **National Union of Women Workers** set up a central committee on the citizenship of women, on the initiative of Mrs Ogilvie Gordon, which, in April 1918, took the name of the National Women Citizens' Association; its first secretary was Helena Normanton. In addition to Women Citizens' Associations formed earlier, a number of members of local suffrage organisations now transferred their concerns to promoting women's citizenship, and either changed the name and objectives of the local suffrage societies or established new local Women Citizens' Associations, which affiliated to the National Association. Some former branches of the Women's Local Government Society also affiliated, and when the Society was dissolved in 1925 they officially changed their names to Women Citizens' Associations. Many Associations had difficulties before and during the Second World War, but after the war they experienced a period of revival, especially in local activities connected with the Festival of Britain in 1951. In 1947 the National Association amalgamated with the **National Council for Equal Citizenship**, and in 1949 with Women for Westminster (campaigning to return women MPs). New Associations were still being formed as late as the 1960s. By the 1950s the branches were organised in five regional federations.

National Women's Register *see* **National Housewives' Register**

National Women's Social and Political Union *see* Women's Social and Political Union

Needlewomen's Institution

This Institution was formed in 1850 to help the underpaid and overworked dressmakers.

New Constitutional Society for Women's Suffrage

Founded in 1910, its aim was 'to unite all suffragists who believe in the anti-government election policy, who desire to work by constitutional means, and to abstain from public criticism of other suffragists whose conscience leads them to adopt different methods'. It appears to have ceased in 1918.

(A. J. R., 1913)

New County Club

Begun in 1899 as the Tea and Shopping Club at 21 Hanover Square, London, 'to provide for ladies of social position the comforts and convenience that men have found in their clubs for years past'. It provided reciprocal membership to the Ladies' Club on the Boulevard Malesherbes, Paris. An orchestra played each afternoon and there was a smoking room. The Club later moved to 138 Piccadilly but closed in 1924.

New Era Club

Founded in 1901 and housed at 110 Victoria Street, London, its main feature was its circulating library to which members could subscribe. Its objects were to promote artistic and literary values and to provide a comfortable meeting place for like-minded spirits. Monday 'At Homes' were held to popularise the Club. It was affiliated to the **Glasgow Society of Lady Artists' Club,** and the Alexandra, Dublin.

New Somerville Club

Founded in 1888 at 231 Oxford Street, London, a year after the closure of the original **Somerville Club**, this club was aimed at ladies living in the suburbs and was housed in a large and more fashionable premises at 19a Hanover Square. It claimed to house a restaurant with a high-class cuisine. The Club mounted a series of debates and lectures on Tuesday evenings on political, social and literary issues and every fourth meeting was a social evening.

New Victorian Club

Established as the Victorian Club for Gentlewomen in 1893 at 30a

Sackville Street, Piccadilly, it was affiliated to clubs in Dublin, St Andrew's, Glasgow and Paris. The facilities were very modest but included accommodation for professional ladies. Tuesday evenings were devoted to bridge and lectures. The Club existed until at least the second decade of the twentieth century.

North of England Council for Promoting the Higher Education of Women

In 1862, Ann Jemima Clough, a trained teacher, made the acquaintance of Barbara Bodichon, Emily Davies and Frances Buss and together with 21 other ladies, decided to form a Schoolmistresses' Association in order to break down the isolation of female teachers. Miss Clough, in an article in *Macmillan's Magazine* the same year, put forward a suggestion of establishing a special Board to regulate the education of girls and arrange lectures in different parts of the country by eminent lectures for older girls and women. In Liverpool, she met Josephine Butler, whose husband was principal of Liverpool College, and the two women agreed to organise women's education in the North of England by promoting a systematic and regular series of university lectures for women, mainly schoolmistresses. Courses of lectures were repeated in a cluster of towns in the same region so that the same lecturer could cover a wide area. A North of England Council for Promoting the Higher Education of Women was formed to administer the scheme with Mrs Butler as president and Miss Clough as secretary. James Stuart, a young Cambridge don, gave lectures on astronomy to women in the autumn of 1867 in Manchester, Liverpool, Sheffield and Leeds. It set the pattern for later lectures, where suitable books were recommended before the lectures and circulating libraries were built up. This developed rapidly; Cambridge University created university extension courses in 1874. By 1870, 25 lecture centres had been established and in 1874, despite reservations both as to the institutional sexism of the university system and the frozen timelessness of Oxbridge syllabuses, it was eventually merged into the newly formed university extension movement.

(Lemoine, 1968; Strachey, 1928)

Office of Lady Helps

Formed in London in the 1870s by Rose Mary Crawshay, the Office sought to secure employment for middle-class women in service as cooks and maids.

Open Door Council

Founded in May 1926 under the leadership of Lady Rhondda, its object was 'to secure that a woman shall be free to work and protected as a worker on the same terms as a man, and that legislation and regulation dealing with conditions and hours, payment, entry and training shall be based upon the nature of the work and not upon the sex of the worker: and to secure for a woman, irrespective of marriage or childbirth, the right at all times to decide whether or not she shall engage in paid work, and to ensure that no legislation or regulation shall deprive her of this right'. In brief, it had the aim of opposing 'protective legislation' which effectively disbarred women, but not men, from the better-paid jobs in industry, e.g. mining; this it termed 'restrictive legislation'. In Britain its membership, which included Elizabeth Abbott and Emmeline Pethick-Lawrence, overlapped considerably with that of the **Six Point Group**, whose outlook it largely shared. Its international equivalent, established in 1929, was the Open Door International. The Open Door Council continued until 1965.

Over Forty Association for Women Workers

Founded in 1934 to provide housing accommodation for working women over the age of 40 who were living on their own in London. Large houses were converted into flats which were let at reasonable rents. Advice on employment was also offered by the Association.

Over Thirty Association

In the 1930s during the economic depression it was calculated that 60 per cent of women registering in Labour Exchanges were over the age of 30. Because of this difficulty in finding suitable employment for this category, the Over Thirty Association was established in 1934 'in the interest of the older industrial professional women workers'. In 1945 it changed its name to the **Over Forty Association for Women Workers**.

Parochial Mission Women's Association

Started in London in 1860 with its headquarters at 11 Buckingham Street, Strand, by four ladies who wished to found an organisation similar to that founded about the same time by Nonconformists. Mission women, selected from people living in the locality, were employed in parochial mission work. They promoted thrift, decency of living, cleanliness and religious belief and collected weekly pence from households which might otherwise be spent unwisely on drink and other 'unfortunate' goods. The money was then used for buying clothing or household commodities. The mission women were poorly paid. The Association was governed by a small committee with a salaried secretary and an assistant. By 1892, the organisation operated in many large towns; in all, 178 mission women were employed.

People's Suffrage Federation

Founded in 1909, this was an 'adult suffrage' society which was at least in theory sympathetic to women's suffrage. It had close links with the Labour movement. Leading members included Mary Macarthur, Margaret Llewelyn Davies and Arthur Henderson.

(A. J. R., 1913; Rover, 1967)

Pinnacle Club

As men's climbing clubs were still closed to women at the beginning of the twentieth century, two women's clubs were founded: the Ladies' Alpine Club (1907) as sister to the Alpine Club and the Ladies' Scottish Climbing Club (1908). A Fell and Rock Climbing Club had been started in 1906 to explore the fells and rock climbs in the Lake District and from the beginning it was open to both men and women. A letter was published in the *Manchester Guardian* in 1920 inviting women who were interested to form their own Society. On 21 March 1921, Mrs Eleanor Winthrop Youngs, who with her husband Geoffrey had been members of the Fell and Rock Climbing Club, took the chair at the inaugural meeting held at Pen y Gwrid Hotel, Llanberis, North Wales. Its initial title was the Women's Rock Climbing Club, but this was soon changed to the Pinnacle Club. Full membership was open to women of proven ability who could lead an ordinary climb of moderate difficulty. The Club aimed to develop a more professional attitude to difficult rock climbing and mountaineering, and develop confidence and good leadership for local expeditions. By the time of its 60th anniversary in 1981, the Society consisted of 150 members, and parties had tackled mountains ranging from the Himalayas to New Zealand.

(Angell, 1988)

Pioneer Club

Started in 1892 by Mrs Emily H. Massingberd, a leading social worker and proponent of temperance, the Club was first located at 180 Regent Street, London,W., over a perfumier's shop, but soon moved, first to 22 Cork Street, London W, then to 22 Bruton Street. The name was a conscious reference to Walt Whitman ('Pioneers! O Pioneers!'), and in 1897 a Frederic Remington sculpture of a 'Pioneer Woman' was presented to the Club. It was open to all classes of society who were united in the common cause of doing good works. It was strongly associated with 'the higher thought' and such associated issues of the 'New Morality' of the late nineteenth century as theosophy, anti-vivisection, anti-vaccination and above all feminism. Activities (which were suspended during August and September) included talks and debates on topical issues. Its Thursday evening debates especially attracted many distinguished speakers such as George Bernard Shaw and Millicent Fawcett, and its activities were reported in the progressive feminist periodical *Shafts*, whose editor, Margaret Shurmer Sibthorpe, was a prominent member. Other members included Eleanor Marx Aveling, Olive Schreiner, Dora Montefiore, Sarah Grand and Mona Caird; the total membership in 1893 was 320. The secretary was Miss Paterson. Membership initially was two guineas per annum, with an entrance fee of one guinea; by 1910 these rates had risen to three guineas with three guineas entrance for town members, and two guineas per annum membership and two guineas entrance fee for country members. Each member wore a small axe, the club's badge, and was known by a number rather than by name, as a symbol of perfect equality. It was the only temperance club in London, though it did provide a smoking room. It also boasted a drawing room, reading room, tea room and visitors' room, as well as bedrooms for the use of members.

At one stage, evening attire for members consisted of a black satin jacket and a white collar and tie rather than more conventional dresses. Although it was reputed initially to have had a sign declaring that 'Gentlemen and dogs are not allowed beyond the vestibule', men were admitted as guests by 1912, if not earlier. In 1897 it was due to move to new premises at 15 Grosvenor Crescent, Hyde Park Corner, but before the move took place Mrs Massingberd died. This seems to have brought to a head internal dissatisfaction with the move, and a substantial proportion of the membership insisted on remaining at the original address and retaining the name of the Pioneer Club. Those who moved took the name of the **Grosvenor Crescent Club**. In fact, the Pioneer Club itself moved shortly afterwards to 5 Grafton Street, Piccadilly, then to 12 Cavendish Place, and then to 9 Park Place, London, SW. It was still active in 1924.

Practical Moral Union of Women of Great Britain and Ireland

This was established in 1833 by a group of women in London led by Mrs Brooks, the aim of which was 'to combine all classes of women' in the struggle for equality. It appears to have ceased activity in 1834.

(Taylor, 1983)

Primrose League

The 1883 Electoral Corrupt Practices Act limited the number of paid political helpers which a parliamentary candidate was allowed at election time. In addition, the Liberal Party had led the way in mounting effective election campaigns by the use of volunteers. To meet this double threat, a small group of Conservative politicians, consisting of Sir Henry Drummond Wolff, Lord Randolph Churchill, Sir John Gorst and Sir Alfred

THE WOMAN ENGINEER

THE ORGAN OF THE WOMEN'S ENGINEERING SOCIETY (Incorporated 1920).

VOL. I., NO. 6. MARCH, 1921. PRICE 6D.

ATALANTA LTD.

THE "WOMAN ENGINEER" IS ISSUED QUARTERLY—PRICE 6d.

1. The organ of the Women's Engineering Society, 1923

THE WOMAN ENGINEER

The Organ of The Women's Engineering Society (Incorporated 1920)

Vol. III. No. 19. JUNE, 1934. Price 6d.

2. The organ of the Women's Engineering Society, 1934

TO THE

WOMEN OF COLCHESTER.

I request you to read the following Extracts from a Parliamentary Report.

They are the words of **Sir Henry Storks.** He says,

"It is to be regretted that the women of Regiments [the wives of Soldiers] cannot be inspected."

By this he means that he desires that they should be treated as common prostitutes, and subjected periodically to the most horrible violation of their persons.

He also says,

"Very little benefit will result from the best devised means of prevention until **Prostitution is recognized as a Necessity.**"

Women of Colchester, are you aware what will be the fate of Womanhood, and of the Nations at large, when Sir Henry Storks' recommendations are carried out, that Prostitution should be recognized as a Necessity, and virtuous women outraged by Act of Parliament?

Are you prepared to stand quietly by, and see the man who has expressed these horrible opinions returned to Parliament as the representative of your husbands and brothers?

I remain, your friend,

JOSEPHINE E. BUTLER,

Honorary Secretary of the Ladies' National Association for the Repeal of the Contagious Diseases Acts.

BACON, PRINTER, QUEEN STREET, COLCHESTER.

3. The Ladies' National Association campaigns against the Contagious Diseases Acts, 1870

Men's Political Union for Women's Enfranchisement.

15, BUCKINGHAM STREET, STRAND, W.C.

The Case of William Ball.

Official Brutality on the Increase.

Last December the 21st, Mr. William Ball, a Birmingham man, and a member of the Transport Workers' Federation, was sentenced to two month's imprisonment for breaking a Home Office window, in protest against the sentence passed on another Suffragist.

In gaol he demanded the treatment laid down by Mr. Churchill's regulations for political offenders of this kind. His demand was refused, and he adopted the hunger strike.

Twice a day for five-and-a-half weeks, the authorities, doctors and warders of Pentonville Gaol fed him by force. Still he held out. Visitors, letters, and examination by doctors from outside were refused him. Nothing could shake his determination to claim the rights granted to political prisoners like Dr. Jameson, and Mr. Ginnell, M.P. Not even the collapse of his mind, under the prolonged and repeated torture, broke his will.

On January the 22nd, Mrs. Ball was informed that he was in his usual health. She heard no more until February the 12th, when she received the following brief and inhuman note :—

"I regret to inform you that it is intended to certify your husband insane on Monday next, the 12th inst., and I am applying to a Justice to attend on Monday to make an order for his detention as a pauper lunatic, and investigate the case.

H. M. DAVIES,
Governor."

Could anything be more pitiless, more heartless, more devoid of common humanity? A man is driven mad by torture, and this is how his wife is informed.

A convicted swindler like Mr. Hooley is even allowed to pursue his profession in gaol. A working man, for a merely nominal political offence is tortured into insanity. We who demand the enfranchisement of women, are thus treated under a Government that calls itself Liberal. The Home Secretary, the Prison Commissioners, the Governors of Gaols are the country's servants. We pay them large salaries; unenfranchised women out of their own pockets help to pay them; in the hope that they will maintain justice. And this is the way in which our paid servants fulfil their trust.

But no matter what their treatment may be, the members of the Men's Political Union, like Mr. Ball himself, will not cease from their endeavour to secure full and equal citizenship for the women of our country, and we call upon all the men of the country who care for freedom, to join our Union, and fight without reservation against this or any other Government that is guilty of such atrocious acts.

HENRY W. NEVINSON,
Chairman of Committee.

4. Leaflet issued by the Men's Political Union for Women's Enfranchisement, 1914

VOTES FOR WOMEN.

The Women's Social and Political Union.

Head Office: 4, CLEMENTS INN, STRAND, W.C.

Telegraphic Address: "Wospolu, London." Telephone No. 2724 Holborn (three lines).

Founder and Hon. Secretary—Mrs. PANKHURST.
Joint Hon. Secretary—Mrs. TUKE.
Publishing Office—THE WOMAN'S PRESS, 156, Charing Cross Road.
Bankers—Messrs. BARCLAY & CO., Fleet Street, E.C.
Hon. Treasurer—Mrs. PETHICK LAWRENCE.
Organising Secretary—Miss CHRISTABEL PANKHURST, LL.B.
Newspaper—VOTES FOR WOMEN.
Colours—PURPLE, WHITE, AND GREEN.

(With apologies to Kipling.)

ELECTORS OF GREAT BRITAIN!

There is before you a Constitutional issue greater than any which has been before you in the lifetime of the present generation. You are called upon to decide whether the people of this country, women as well as men—the whole people and not one privileged sex only—shall have the Vote. You are also called upon to decide whether your elected representatives in the House of Commons are to be free to carry out the people's will, unfettered by the **autocratic and irresponsible veto of the Prime Minister and his Cabinet.**

If you believe in political justice for those women who can show the same claim to the Vote as a man elector can show, and if you believe that the House of Commons ought to

5. Leaflet issued by the Women's Social and Political Union, 1910

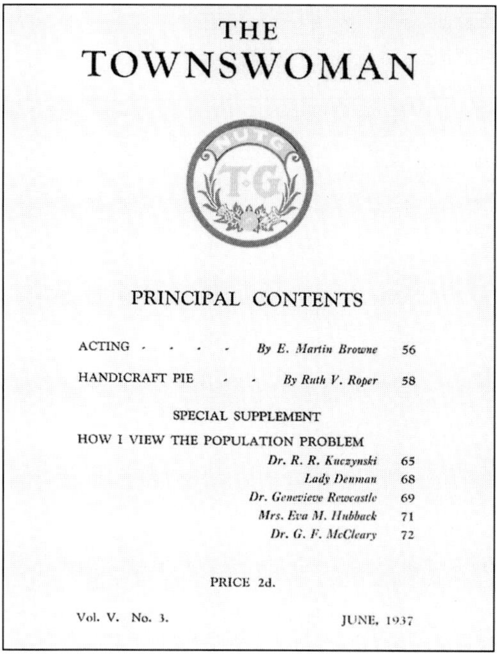
THE
TOWNSWOMAN

PRINCIPAL CONTENTS

ACTING - - - - By E. Martin Browne 56

HANDICRAFT PIE - - By Ruth V. Roper 58

SPECIAL SUPPLEMENT

HOW I VIEW THE POPULATION PROBLEM

Dr. R. R. Kuczynski 65
Lady Denman 68
Dr. Genevieve Rewcastle 69
Mrs. Eva M. Hubback 71
Dr. G. F. McCleary 72

PRICE 2d.

Vol. V. No. 3. JUNE, 1937

6. The organ of the Townswomen's Guilds, 1937

7. The Townswomen's Guilds give advertising space to a government campaign, 1937

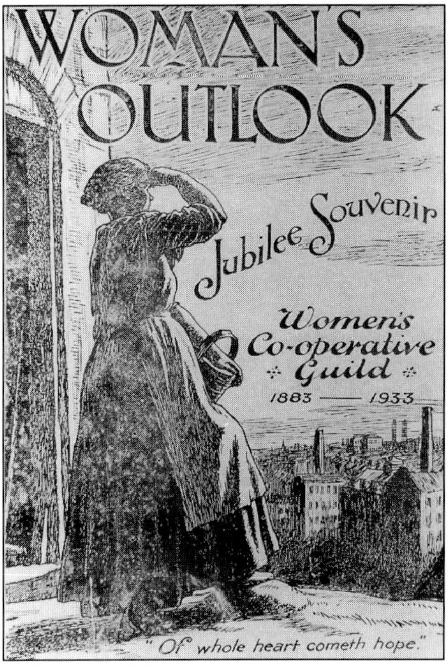

8. The Women's Co-operative Guild celebrates its Jubilee, 1933

9. The *Women's Voluntary Service Bulletin*, 1951

10. The Women's Voluntary Service celebrates the Festival of Britain, 1951

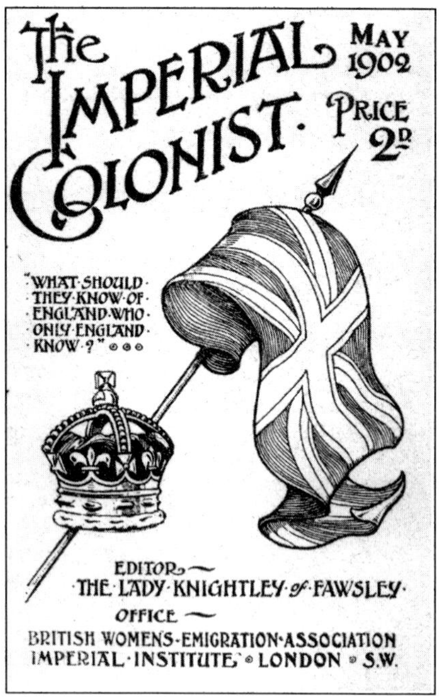

11. Reporting activities of the British Women's Emigration Association, 1911

12. The organ of the National Council of Women, 1934

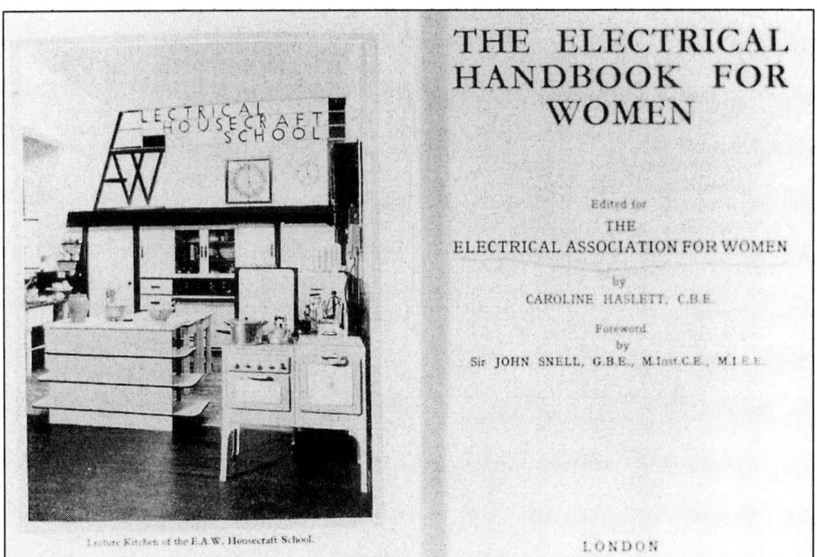

13. The Electrical Association for Women publishes a handbook, 1934

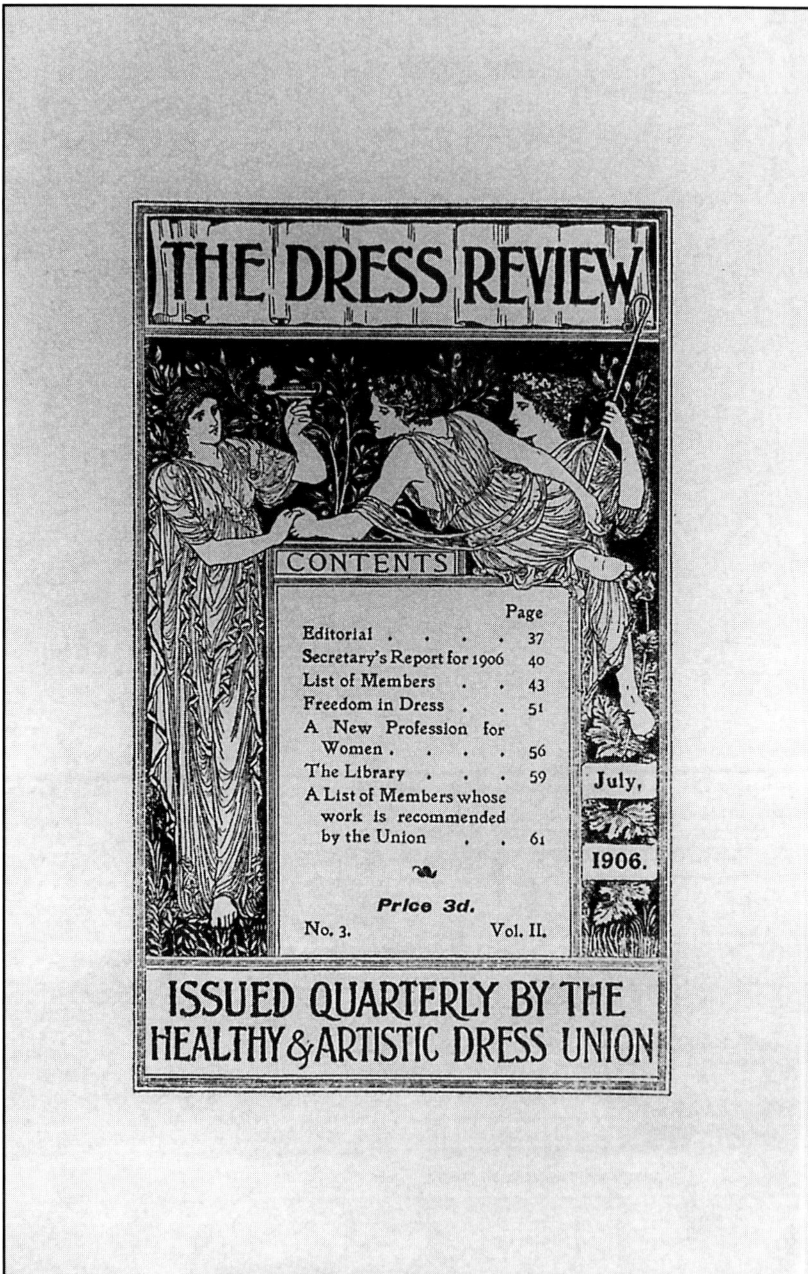

THE DRESS REVIEW

CONTENTS

Price 3d.

No. 3. Vol. II.

July, 1906.

ISSUED QUARTERLY BY THE HEALTHY & ARTISTIC DRESS UNION

14. The organ of the Healthy and Artistic Dress Union, 1906

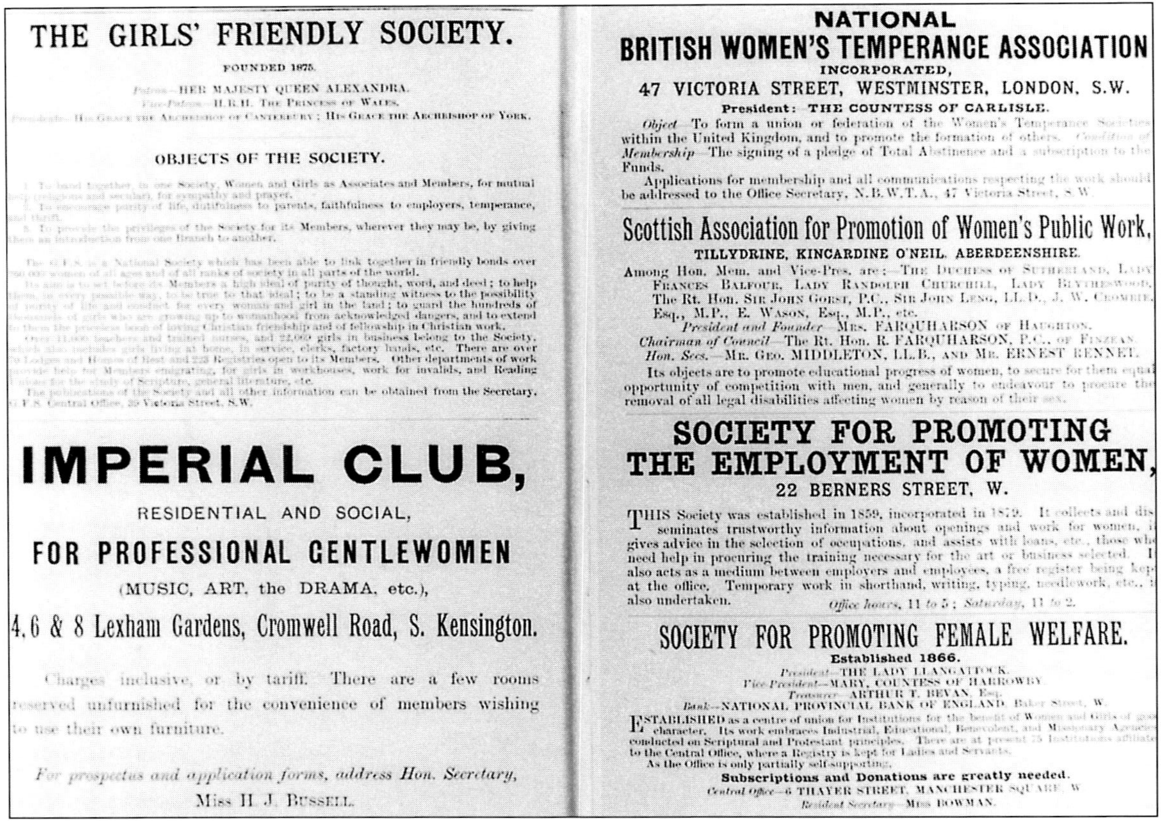

THE GIRLS' FRIENDLY SOCIETY.

FOUNDED 1875.

Patron—HER MAJESTY QUEEN ALEXANDRA.
Vice-Patron—H.R.H. THE PRINCESS OF WALES.
Presidents—HIS GRACE THE ARCHBISHOP OF CANTERBURY; HIS GRACE THE ARCHBISHOP OF YORK.

OBJECTS OF THE SOCIETY.

1. To band together, in one Society, Women and Girls as Associates and Members, for mutual help (religious and secular), for sympathy and prayer.
2. To encourage purity of life, dutifulness to parents, faithfulness to employers, temperance, and thrift.
3. To provide the privileges of the Society for its Members, wherever they may be, by giving them an introduction from one Branch to another.

The G.F.S. is a National Society which has been able to link together in friendly bonds over 200,000 women of all ages and of all ranks of society in all parts of the world.

Its aim is to set before its Members a high ideal of purity of thought, word, and deed; to help them, in every possible way, to be true to that ideal; to be a standing witness to the possibility of purity of life and conduct for every woman and girl in the land; to guard the hundreds of thousands of girls who are growing up to womanhood from acknowledged dangers, and to extend to them the priceless boon of loving Christian friendship and of fellowship in Christian work.

Over 11,000 teachers and trained nurses, and 22,000 girls in business belong to the Society, which also includes girls living at home, in service, clerks, factory hands, etc. There are over 70 Lodges and Homes of Rest and 223 Registries open to its Members. Other departments of work provide help for Members emigrating, for girls in workhouses, work for invalids, and Reading Unions for the study of Scripture, general literature, etc.

The publications of the Society and all other information can be obtained from the Secretary, G.F.S. Central Office, 39 Victoria Street, S.W.

IMPERIAL CLUB,

RESIDENTIAL AND SOCIAL,

FOR PROFESSIONAL GENTLEWOMEN

(MUSIC, ART, the DRAMA, etc.),

4, 6 & 8 Lexham Gardens, Cromwell Road, S. Kensington.

Charges inclusive, or by tariff. There are a few rooms reserved unfurnished for the convenience of members wishing to use their own furniture.

For prospectus and application forms, address Hon. Secretary,
Miss H. J. BUSSELL.

NATIONAL
BRITISH WOMEN'S TEMPERANCE ASSOCIATION
INCORPORATED,
47 VICTORIA STREET, WESTMINSTER, LONDON, S.W.

President: THE COUNTESS OF CARLISLE.

Object—To form a union or federation of the Women's Temperance Societies within the United Kingdom, and to promote the formation of others. *Condition of Membership*—The signing of a pledge of Total Abstinence and a subscription to the Funds.

Applications for membership and all communications respecting the work should be addressed to the Office Secretary, N.B.W.T.A., 47 Victoria Street, S.W.

Scottish Association for Promotion of Women's Public Work,

TILLYDRINE, KINCARDINE O'NEIL, ABERDEENSHIRE.

Among Hon. Mem. and Vice-Pres. are:—THE DUCHESS OF SUTHERLAND, LADY FRANCES BALFOUR, LADY RANDOLPH CHURCHILL, LADY BLYTHESWOOD, The Rt. Hon. SIR JOHN GORST, P.C., SIR JOHN LENG, LL.D., J. W. CROMBIE, Esq., M.P., E. WASON, Esq., M.P., etc.

President and Founder—MRS. FARQUHARSON OF HAUGHTON.
Chairman of Council—The Rt. Hon. R. FARQUHARSON, P.C., OF FINZEAN.
Hon. Secs.—MR. GEO. MIDDLETON, LL.B., AND MR. ERNEST BENNET.

Its objects are to promote educational progress of women, to secure for them equal opportunity of competition with men, and generally to endeavour to procure the removal of all legal disabilities affecting women by reason of their sex.

SOCIETY FOR PROMOTING
THE EMPLOYMENT OF WOMEN,
22 BERNERS STREET, W.

THIS Society was established in 1859, incorporated in 1879. It collects and disseminates trustworthy information about openings and work for women, it gives advice in the selection of occupations, and assists with loans, etc., those who need help in procuring the training necessary for the art or business selected. It also acts as a medium between employers and employees, a free register being kept at the office. Temporary work in shorthand, writing, typing, needlework, etc., is also undertaken.

Office hours, 11 *to* 5; *Saturday*, 11 *to* 2.

SOCIETY FOR PROMOTING FEMALE WELFARE.

Established 1866.

President—THE LADY LLANGATTOCK.
Vice-President—MARY, COUNTESS OF HARROWBY.
Treasurer—ARTHUR T. BEVAN, Esq.
Bank—NATIONAL PROVINCIAL BANK OF ENGLAND, Baker Street, W.

ESTABLISHED as a centre of union for Institutions for the benefit of Women and Girls of good character. Its work embraces Industrial, Educational, Benevolent, and Missionary Agencies conducted on Scriptural and Protestant principles. There are at present 75 Institutions affiliated to the Central Office, where a Registry is kept for Ladies and Servants.

As the Office is only partially self-supporting,

Subscriptions and Donations are greatly needed.

Central Office—6 THAYER STREET, MANCHESTER SQUARE, W.
Resident Secretary—MISS BOWMAN.

15. Women's organisations advertise in the *Englishwoman's Yearbook*, 1910

MARY META BAGOT STACK
(1883–1935)
THE FOUNDER OF THE LEAGUE

MOVEMENT IS LIFE

The Intimate History of the Founder of the
WOMEN'S LEAGUE OF HEALTH AND BEAUTY
and of its Origin, Growth, Achievements,
and Hopes for the Future

BY

MRS. A. J. CRUICKSHANK
AND
PRUNELLA STACK
Joint Leaders of the League since 1935

G. BELL AND SONS, LTD
LONDON 1937

16. Mary Bagot Stack, founder of the Women's League of Health and Beauty, and the League's logo, 1937

17. The Six Point Group commemorates Dorothy Evans, a prominent member, 1946

The Society for the Oversea Settlement of British Women (Incorporated).

CONSTITUTED 1919 FROM

The British Women's Emigration Association. (Founded 1884).
The South African Colonization Society. (Founded 1902).
The Colonial Intelligence League. (Founded 1911).

COUNCIL.

Chairman: The DOWAGER VISCOUNTESS HARCOURT, G.B.E.

Vice-Chairmen { The Hon. Lady Cecil, C.B.E.
The Hon. Mrs. Wilson Fox, C.B.E.

*Miss Balfe (*Catholic Women's League*).
Mrs. Harrison Bell.
Mrs. Moberly Bell.
Mrs. Black.
The Countess Buxton, G.B.E.
The Hon. Mrs. Francis Buxton, O.B.E.
Lady Campion.
The Countess of Clarendon.
Miss H. F. Cohen, O.B.E.
Lady Davson, O.B.E.
Mrs. Geoffrey Dawson.
Miss R. F. Faudel-Phillips.
*The Lady Forster, G.B.E. (*National Council of Women*).
Miss A. C. Franklin, O.B.E.
Lady Furse, O.B.E.
Lady Goold-Adams.
The Hon. Mrs. Norman Grosvenor, C.B.E.
Miss O. M. Hargreaves.
Mrs. Hilder.
The Hon. Mrs. Kitson.
Mrs. Herbert Lane.
*Mrs. Moore (*Standing Joint Committee of Industrial Women's Organisations*).
Dame Beryl Oliver, D.B.E., R.R.C.
*Miss Addison Phillips (*Association of Head Mistresses*).
Miss E. Ransford.
Lady Reed.
Miss Violet Russell.
*Mrs. de Satgé (*Mothers' Union*).
Miss F. Saunders.
Miss W. G. Scarth.
Mrs. Simm.
Lady Solomon.
The Dowager Lady Stanley of Alderley.
Dame Meriel Talbot, D.B.E.
*Miss Helen Talbot (*Girl Guides*).
Miss Edith Thompson, C.B.E.
Miss J. Varley, O.B.E.
*Miss Warman (*G.F.S.*).
Mrs. G. Whiskard.
Miss M. Wilde, M.B.E.
*Miss Wood (*Y.W.C.A.*)

* These members represent the Society indicated in brackets after their name.

EXECUTIVE COMMITTEE.

Chairman: Miss Gladys Pott, O.B.E.

Mrs. Harrison Bell.
Mrs. Black.
The Countess Buxton, G.B.E.
The Hon. Lady Cecil, C.B.E.
Lady Davson, O.B.E.
The Hon. Mrs. Wilson Fox, C.B.E.
Lady Furse, O.B.E.
The Hon. Mrs. Norman Grosvenor, C.B.E.
Miss O. M. Hargreaves.
The Hon. Mrs. Kitson.
Dame Beryl Oliver, D.B.E., R.R.C.
Miss Violet Russell.
Lady Solomon.
Miss Edith Thompson, C.B.E.
Miss Warman.

Medical Officer: Dr. Ethel Vernon, M.D., B.S.(Lon.), 23, Grosvenor Road, S.W.1.
Hon. Treasurers: W. Bankes Amery, Esq., C.B.E., and Major C. S. Goldman.
Bankers: Westminster Bank, Ltd.
Hon. Organiser: Miss A. C. Franklin, O.B.E.
Secretary: Miss E. S. Nicholas.

Interviews: 11 to 4, Saturdays 11 to 12-30. TELEPHONE: Victoria 3730.
TELEGRAMS: Mitigator, Sowest, London. CABLES: Mitigator, London.

54, VICTORIA STREET, WESTMINSTER, S.W. 1.
(Temporary address until June).

18. Page from the report of the Society for the Oversea Settlement of British Women, 1920

19. Report of the National Federation of Women's Institutes, 1923

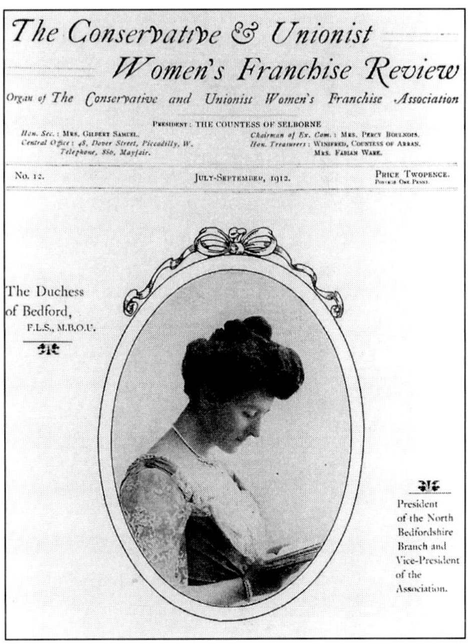

20. The organ of the Conservative and Unionist Women's Franchise Association, 1912

The Labour Woman
Edited by Dr. Marion Phillips
Vol. XVI No. 4
Price Twopence
Subscription with Postage Three Shillings Annually
April, 1928
PUBLISHED BY THE LABOUR PARTY
EQUAL FRANCHISE
WELCOME TO THE NEW WOMAN VOTER
A Monthly Journal for Working Women

21. Labour women celebrate the Equal Franchise Act, 1928

THE LAND GIRL

No. 8. Vol. I. NOVEMBER 1940 Price 3d.

QUIET GALLANTRY

At our breakfast tables we drink our milk and eat our bread and eggs or bacon (when we can get them), and perhaps we sometimes think warmly of the Land Army which has helped to bring them to us, but how often do we think of the courage in the battle line, of which our breakfasts are the result?

At half-past five one morning recently a bomb fell on a road in the outskirts of a large town. Twenty-five yards away a Land Girl, on her round, was flung from road to pavement and her pony gravely injured. Dazed but unhurt, she managed to get a message to her employer farmer, waited with the pony until he and the vet. arrived, and then went on with her round on foot.

Read in the County News of the Dorset girls, bombed both at home and at work, who went on with their milk bottling. Read also of Kent volunteers. Here over 200 have worked on steadily in constant danger. Six of them have been doing tractor ploughing in a particularly dangerous coastal barrage area. It is not " done " to shelter until the battle is overhead—otherwise one would be for ever sheltering. But the tractors were so noisy that the drivers could not tell when the battle *was* overhead. So frequently they found themselves working with shrapnel falling all round them and even on their tractors. Not one of these girls asked for a transfer. Instead they applied for tin hats—and got them—and went on ploughing. The East Kent Secretary has been bombed twice in her own house and once where she was staying; her office is in a town where houses fell like packs of cards around her, but she worked there every day and could not be persuaded to move. In East Sussex a Land Girl was the first to arrive to rescue two small children from a cottage near which a burning aeroplane had crashed.

It is courage like this, repeated again and again in every county and every town, which cannot be beaten, for it is incapable of acknowledging defeat. The Land Army is proud that its members are justifying so magnificently our confidence in victory.

M. A. P.

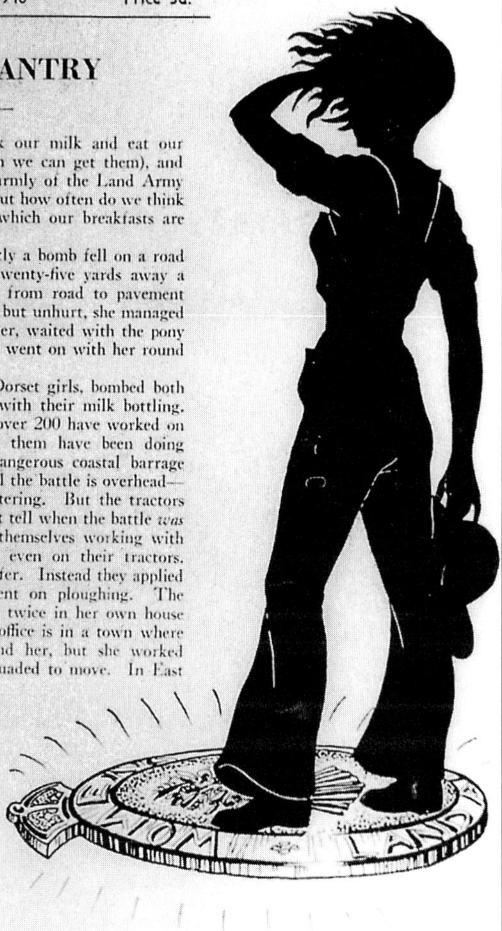

22. News for the Women's Land Army, 1940

23. A concert at the Jewish Working Girls' Club, Soho, 1901

24. A debate at the Pioneer Club, 1902

Slade, conceived the idea of forming a Primrose Tory League. The League, which took as its emblem Disraeli's favourite flower, was to 'embrace all classes and creeds except atheists and enemies of the British Empire', and members were to subscribe to 'the maintenance of religion, and of the Constitution of the realm, and of the Imperial ascendancy of Great Britain'. The League encouraged the involvement of like-minded people in a district into groups called habitations, which would become active bodies during election campaigns. Almost from the beginning women were allowed to join as honorary members; on the payment of a fee, men were admitted as Knights and women as Dames. The League had as its head a Grand Master and a Grand Council; these offices were not open to women.

In March 1885, Lady Glenesk, wife of the proprietor of the *Morning Post*, together with 14 other women, formed a Ladies' Grand Council (LGC) with a membership fee of a guinea a year. An Executive Committee of 15 members was formed which rotated every three years. The LGC was active in financing and distributing political leaflets, sending out vans with magic lanterns to rural areas, and in providing speakers for rallies. However, the main success of the League was in recruiting women to work locally in habitations, sometimes forming exclusively female habitations. By the end of the century, the League consisted of almost 1.5 million members. Their main function was to canvass at election time, to supervise registration and to organise fêtes, which consisted of a mixture of entertainments and political propaganda, and other activities. The army of women workers made an immediate impact on the outcome of the 1885 general election and proved to be an important factor in subsequent campaigns. News of the League's activities was published in the *Primrose League Gazette*. An attempt to produce a journal with a more popular appeal, the *Primrose Herald*, was not a success. On the question of women's suffrage, the League held no official opinion, but in 1889, the Grand Council forbade habitations taking part in suffrage meetings or passing resolutions in support of the franchise for women. The League declined in numbers and influence after women got the vote in 1918 and other competing organisations were set up by the Conservative Party.

(Robb, 1942)

Princess Mary's Royal Air Force Nursing Service

The Royal Air Force Nursing Service was formed in 1918 and was given its present title in 1923. It recruited qualified nurses for service hospitals at home and overseas and during the Second World War served in every theatre of war. Until March 1943, Sisters, though enjoying the status of officers, were not commissioned. Now, commissions are available for Registered General Nurses who have at least two years' experience.

Queen Alexandra's Royal Army Nursing Corps

The history of Army nursing started with Florence Nightingale's mission during the Crimean War in 1854 when 40 nurses were sent to Russia. Nurses worked in military hospitals until 1881 when the Army Nursing Service was inaugurated. From 1889, Sisters were employed in all Army hospitals with 100 or more beds. In 1902, the title of the Service was changed to the Queen Alexandra's Imperial Military Nursing Service. By the end of the Second World War, when the Service had expanded to cover all the theatres of war, there were 10,404 trained nurses. The present title, the Queen Alexandra's Royal Army Nursing Corps, was adopted in 1949. QARANC has trained nurses from 1950. Since 1992, men have been eligible to join.

(*Nursing in the Army,* 1905; Wadge, 1946)

Queen Alexandra's Royal Naval Nursing Service

The Royal Naval Nursing Service was formed in 1884 with the appointment of Nursing Sisters, at a salary of £25 per annum, to the Naval Hospital at Haslow. Trained sick bay staff were also recruited. In 1902, Queen Alexandra became the Service's president and its title was changed. Members did not hold commissions nor were they subject to naval discipline. The QARNNS was expanded during the Second World War. Men were integrated into the Service from 1987.

Queen Mary's Army Auxiliary Corps *see* Women's Army Auxiliary Corps

Queen Mary's Committee for Women's Training and Employment

The Committee was set up 12 days after the outbreak of the First World War in August 1914 to assist unemployed women by providing training schemes and work. Queen Mary, as Princess of Wales, had visited the Sweated Trade Exhibition in 1906 and she became patroness of the Committee when it was formed. Committee members consisted of women of all political shades, ranging from the chairman, Lady Crewe, to Mary Macarthur. After the war, Macarthur became chairman and in 1920 persuaded the Ministry of Labour to allow the Committee to be responsible for the organisation and supervision of training schemes for women who had suffered earning losses through the war and to help redundant women workers find jobs.

(Lewenhak, 1977)

Queen's Club

Founded in 1897 at 7 Frederick Street, Edinburgh as a social meeting place, the Club was limited in membership to 500. Nomination was by two existing members and election by ballot. Management was by a board of trustees. Short-term accommodation was provided for members. The entrance fee was five guineas, with a subscription for town members five guineas and country members four guineas. The secretary was Miss Lena Meiklejohn.

Queen's Nursing Institute

To mark the Golden Jubilee of Queen Victoria's reign in 1887, a Jubilee Fund, organised by 'the women of England' raised £70,000. The Queen announced that the offering should go to providing nursing. As a result, this Institute, initially called the Queen Victoria's Jubilee Institute for Nurses, was established by Royal Charter in 1889. Its aims were to train district nurses and to supervise county and district nursing associations. The training period was three years with those employed in midwifery receiving additional training. Successful candidates were known as Queen's Nurses. Rosalind Paget, niece of William Rathbone, was appointed as its first Inspector-General. In 1889, he was one of the prime movers in founding the training school for its nurses at the **Metropolitan and National Nursing Association** in Bloomsbury Square, London. Queen Alexandra agreed to be patron of the Institute in 1901. By 1916, there were 2,000 Queen's Nurses working in 1,000 affiliated associations. In 1928 it became the Queen's Institute of District Nursing and in 1972 it assumed its present title, the Queen's Nursing Institute.

(Baly, 1987)

'Qui Vive' Corps

Founded in 1913 by Florence de Fonblanque and Ruth Cavendish-Bentinck, this was intended to be a sort of suffragist 'flying squad', with its own uniform, brown with green cockade and badge; members had 'undertaken to turn out at any moment and rally at any given spot'; however, they were pledged to abstain from any militant action. 'On the qui vive', in the idiom of the time, meant 'On the alert'. Its activities in practice seem to have been mainly concentrated on Sussex.

(A. J. R., 1913)

Rational Dress Society

It was formed in 1888 to protest 'against the introduction of any fashion in dress that either deforms the figure, impedes the movement of the body, or in any way tends to injure the health'. Particularly objected to were tight-fitting corsets, high-heeled and narrow-tied boots and shoes, heavily weighted skirts, crinolines and crinolinettes. Its maximum approved weight of women's clothing was seven pounds, approximately 3.2 kilogrammes, without shoes. Its original secretary was Mrs Carpenter-Fenton, later succeeded by Mrs Hall, and its first address was 23 Mortimer Street, London, W. Later it moved to 11 Sloane Street, London, SW. The annual subscription was 3s 6d, and its (apparently short-lived) organ was the *Rational Dress Society's Gazette*.

Rehearsal Club

Founded in 1892 at 29 Leicester Square, London, in consequence of the increasing popularity of the theatre, the Club provided a homely centre for minor actresses, higher grade members of the corps de ballet and members of the chorus in various music halls between morning and evening performances. Tea and light refreshments were available. The Club was the brainchild of the Rev. John Fenwick Kitto, vicar of St Martin-in-the-Fields, who also established a similar venture, the **Wantage Club**, for waitresses working at hotels and restaurants.

Royal British Legion: Women's Section

This organisation was established in 1921 to promote the welfare of ex-Service women, wives, widows and children and dependants of those who had served in the Armed Forces and were in need of help. It is also active in trying to influence government policy on ex-Service people. The Section is involved in assisting the Royal British Legion by raising funds. It worked through local branches and besides providing flatlets and rest homes, paid visits to in-patient and war widows living at home, caring for the welfare of disabled ex-servicemen and collecting for the Legion's Poppy Day appeal. The Section is open to all over the age of 16.

Royal British Nurses' Association

The Association arose from an organisation of hospital administrators, the Hospitals' Association, which set up a committee in 1886 to urge the registration of nurses after a one-year course of training. However, a meeting attended by 30 matrons, led by Ethel Bedford Fenwick, at St Bartholomew's Hospital, London, concluded that a three-year training course was needed. In November 1887, a conference of matrons formed the British Nursing Council, 'a union of nurses for professional objects', with Mrs Fenwick as its permanent president and the Association was formally launched in February 1888. It became a pressure group for state registration. On 1 January 1889, it admitted to its register those who could demonstrate a satisfactory standard of nursing as well as having at least three years' experience. The Association also hoped that national standards could be laid down and that every hospital would form a board to organise nurse education, though results were disappointing in this aspect. In 1891, it was granted a Royal Charter, becoming the Royal British Nurses' Association, with Princess Christian as its patron.

(Abel-Smith, 1960)

Royal College of Midwives

In 1869 a survey by the Obstetrical Society of London had shown that the majority of midwives were uneducated women from the poorer classes. Three years later the Society instituted an examination for midwives but it was not a mandatory requirement. As a result, a group of well-qualified midwives under the leadership of Louisa Hubbard and Zepharina Veitch (Mrs Harry Smith), formed the Matrons' Aid Society, which became in 1881 the Midwives' Institute. Its objects were to raise the efficiency and improve the status of midwives, to establish a register of members and open an information centre for the public. It established a good medical library and a club room for members, and arranged courses of medical lectures.

The Institute was incorporated in 1889 and a Midwives' Registration Bill was brought before Parliament the following year. It encountered opposition in the Commons as did similar attempts later in the decade. The Institute acted as an effective pressure group and petitioned Parliament for recognition; this led to the first Midwives Act of 1902 to 'secure better training of midwives and to regulate their practice'. A Central Midwives Board was established in England, which *inter alia* governed the conditions for entry into the profession. Unqualified midwives were forbidden to practice. However, it was not until the Midwives Act of 1936 that unqualified women were successfully prohibited from attending at childbirth. In 1941 the Midwives Institute became the College of Midwives and was given the prefix 'Royal' in 1947. The College represents the rights of midwives to legal and political authorities and reports on and encourages research. The 1979 Nurses, Midwives and Health Visitors Act dissolved all statutory and non-statutory training bodies. The Central Midwives Board was replaced by a Central Council for the three services.

(Hart, 1994)

Royal College of Nursing

During the First World War, it became obvious that trained nurses needed their own democratic organisation on a national scale. The College of Nursing was registered as a limited company on 1 April 1916, with a number of objectives: to promote better education and training, to encourage uniformity of curriculum, to recognise approved nursing schools, to maintain a register of proficient nurses and to promote bills in Parliament. By 1913, there were over 13,000 members and the **Royal British Nurses' Association** was approached with a view to amalgamation but this was not successful. In 1919, a Nurses' Bill received the Royal Assent which established the General Nursing Council. Of the 15 members of the Council, the College had nine seats, the RBNA four seats and the Poor Law Institutes two. Its main duty was to set up a register of nurses and compile a syllabus of subjects for examination. The first newly qualified nurses, called State Registered Nurses, completed their training in 1922. Now with 20,000 members, the College expanded its role to include continuing education for nurses and it campaigned for improved pay and conditions of service. It was given the title 'Royal' by George VI in 1939.

(Bowman, 1967)

Royal Female Philanthropic Society

It is mentioned in the *Englishwoman's Year Book* for 1910, with an address at Manor Hall, 53 Great Church Lane, Hammersmith, London, W.

Royal Irish Association for Promoting the Training and Employment of Women

This Association was established in 1882 for the purpose of providing technical training for women in order to equip them for employment. The Association also undertook to secure work for them. Instruction was given in many subjects including scrivenery, plain tracing, wood carving, illumination, typewriting and shorthand. Students were given two lessons a week at the cost of five shillings. The prefix 'Royal' was granted in 1888 when Queen Victoria became its patroness. Up to 1885, the Association operated mainly in Dublin, but in that year it was resolved to extend its operations to the whole of Ireland and the Association was reorganised on this basis.

Saint Andrew's House Club

A proprietary Club, run by Miss Edith Debenham, for women engaged in nursing or other professions, it was located at Mortimer Street, London, W. Its objectives were social and it provided residential accommodation. Qualification was through election by a committee. The entrance fee was one guinea with country members paying a subscription annually of half a guinea and town members, one guinea. In 1906 there were 400 members.

Saint Joan's Social and Political Alliance

Continuing the **Catholic Women's Suffrage Society**, from 1923 it has served as a campaigning focus for Roman Catholic feminists. Its main focus has always been more feminist than strictly Catholic, since it has actively supported campaigns over such issues as the nationality of married women, equal pay and reform of laws relating to prostitution; more recently it has confronted the issue of women's ordination. However, in the 1920s and 1930s it often found itself at variance both with most other feminists over such issues as birth control, while also being at odds with orthodox Catholic views of women. In 1961 it changed its name to the Saint Joan's Alliance. Its organ is the *Catholic Citizen.* Since 1921 it has had its counterpart in the Saint Joan's International Alliance, which is at the time of writing its most visible aspect.

(Parnell, 1961)

Saint Mary's Club for Working Girls *see* Women's University Settlement

Saint Monica's Home of Rest *see* Women's University Settlement

Salvation Army Home League

Founded in 1907, the League was open to women over the age of 16. It provided for education, fellowship, service and worship.

Sandringham Club

Established in 1899 at 38 Dover Street, Piccadilly, with a strong emphasis on sporting facilities. Golfing was a popular feature, with an annual competition for the Sandringham Golf Cup. A billiards tournament was an equally popular event: a lady professional was in attendance to provide lessons. Residential facilities were available. By 1900 there were 400 members.

School Nurses' League

Earlier organisations in the last decade of the nineteenth century, such as the Queen Victoria's Jubilee Institute for Nurses, later the **Queen's Nursing Institute**, and the **Metropolitan and National Nursing Association**, had recognised the need to attend to the health of the school child. In February 1908, Mrs Bedford Fenwick established the School Nurses' League. The nucleus of members at first consisted of London County Council nurses but later membership was open to all parts of the United Kingdom.

Scottish Churches' League for Woman Suffrage

Formed in 1912, with Lady Frances Balfour as president, it aimed to bring members of Scottish churches together to campaign as such for women's suffrage, and to promote the cause of women's spiritual and political equality within the churches.

(A. J. R., 1913)

Scottish Co-operative Women's Guild

This Guild was begun in 1892 to bring together all branches of the Guild which existed in order to extend and strengthen the Co-operative movement, locally, nationally and internationally. The Guild raises questions on social and political reform and representation of women in government and supports legislation which will be of benefit to women.

Scottish Council for Women's Trades

In 1895, when the Trades Union Congress decided to exclude trade councils from affiliation, Margaret Irwin, who was the Scottish organiser for the **Women's Trade Union League**, left to become secretary for the Scottish Council for Women's Trades. She was at the same time appointed secretary of the Scottish Trades Union Congress Committee. The Council set out to limit working hours for shop assistants and non-stop shifts in laundries. It later took up the cause of seasonal workers. Because of its membership, the Council adopted a middle-class attitude towards philanthropy. As late as 1929, the Council, in finding work for unemployed women as domestics in farmhouses, vetted the character and qualifications of candidates on behalf of future employers.

Scottish University Women's Suffrage Union

It was founded in 1909 to further the cause of women's suffrage among the Scottish universities and their graduates; its secretary was Chrystal Macmillan.

(A. J. R., 1913)

Scottish Women's Rural Institutes

This organisation was established in 1917 with its headquarters at 42 Heriot Row, Edinburgh. Its objectives are to advance the education and training of those who live and work in the country by greater knowledge of home skills, family welfare and citizenship. The Institutes aim to preserve the traditions of rural Scotland, encourage small businesses and work for international co-operation and understanding amongst women. It issues a monthly magazine *Scottish Home and Country.* Membership is approximately 32,000 women.

Sea Rangers

Founded by Dame Katherine Furse, who had been the first director of the **Women's Royal Naval Service** after its establishment in 1917, and Mrs (later Dame) Vera Laughton Mathews, director of the Service in the Second World War. Furse was a Commissioner of the **Girl Guides** and decided to set up an equivalent organisation, the Sea Guides, later known as the Sea Rangers. Young women were taught rowing, sailing and astronomy as well as how to make knots and lanyards, and were made aware of the tradition of the sea and the Royal Navy. During the Second World War, many former Sea Rangers joined the WRNS.

(Furse, 1940)

Sesame Club

Established in 1895 at 25 Dover Street, London, it aimed at attracting professional women and provided educational talks and discussions. Both men and women were eligible for membership but fees were lower for women. Prominent members included Lady Isabel Margesson, May Sinclair, Margaret Tuke, Lady Betty Balfour and Catherine Marshall. At some point after the mid-1920s it appears to have amalgamated with the **Ladies' Imperial Club** and the **Pioneer Club**. In 1963 its address was 49 Grosvenor Street, London,W1.

Sisters of Charity

Founded by Elizabeth Fry shortly after a visit to Kaiserwerth near Düsseldorf in Germany in 1840, it was the first institution to establish training for nurses and those religious women wishing to nurse the poor. When their training was completed, the nurses then either took posts in hospitals with which they were already familiar, accepted private posts or did district nursing for the poor. In 1841, Queen Adelaide became the Royal patron. From the beginning of the twentieth century, the nurses undertook a three-year initial training in hospitals.

(Jameson, 1855)

Sisters of the People

It was formed in 1887 by Dorothy Price Hughes, wife of the founder of the West London Mission, as a Sisterhood of 'devout and educated women' helping poor women and girls in the Fitzrovia, Soho and Marylebone districts of London. Although the Mission was a Wesleyan Methodist foundation, the Sisterhood did not insist on Methodist credentials for its members. Prominent Sisters included Mary Neal and Emmeline Pethick-Lawrence, and it was the starting point of the **Espérance Girls' Club**.

Six Point Group

This was founded by Lady Rhondda in 1921 to address women's equality issues; in its early years, Lady Rhondda's *Time and Tide* acted as its organ. The six points were: political equality; occupational equality; moral equality; social equality; economic equality and legal equality. It was associated with the 'old feminism' or 'equality feminism' of the 1920s, as distinct from what its members perceived as the 'new' or 'welfare' feminism of the **National Union of Societies for Equal Citizenship**.

However, this opposition was mainly from a 'left' standpoint, as distinct from the more conventionally liberal–economic position of such older feminists as Millicent Fawcett. Many members were former militant suffragists, like Dorothy E. Evans, Monica Whately and Helen Archdale (the first editor of *Time and Tide*), but the Group also attracted many younger women, like Winifred Holtby and Vera Brittain. Its influence in the interwar years and the Second World War was considerable, and out of proportion to its relatively small numbers. It continued to be active until the 1970s, when its secretary was Hazel Hunkins Hallinan, and it finally ceased in 1980.

(Beddoe, 1989; Eoff, 1991)

Slade Club

The object of this Club was to provide a rendezvous for past and present women art students of the Slade School, University College, London. In 1906, it occupied premises at the Constable Studios, 76 Charlotte Street, Fitzroy Square, London. The entrance fee was five shillings, and the annual subscription was one guinea for town members and a half a guinea for country members. The honorary secretary and treasurer was Miss C. Bentinck Smith.

Snowdrop Bands

Popular in the late 1880s and 1890s, the movement aimed to recruit factory girls in towns in the Midlands and the North of England in order 'to purify and exalt femininity'. Miss Nunneley, a member of the Birmingham Ladies' Union of Workers among Women and Girls, urged girls to 'promise, with God's help, to earnestly try, both by our example and influence, to discourage all wrong conversation, light and immodest conduct and the reading of bad and foolish books'. Members were given a card depicting a bunch of snowdrops and inscribed 'White flowers of a blameless life', and round the edge, 'Let thy mind's sweetness have its operation on thy body, clothes and habitation.'

(Dyhouse, 1981)

Society for Constructive Birth Control and Racial Progress

On 17 March 1921, Marie Carmichael Stopes and Humphrey Roe opened the Mothers' Clinic for Constructive Birth Control, the first British birth control clinic, at 61 Marlborough Road, London, N9 (it later moved to 108 Whitfield Street, London, W1). In August 1921, the Society was formed, with Marie Stopes as president. Its main purpose was stated to be not only disseminating birth-control propaganda but recognising the 'sacredness of motherhood'. In theory it had fairly strong eugenical leanings; birth control was to be instrumental in improving the stock of the human race. However, advice given at the clinic did not always strictly follow these principles. The preferred form of contraception was the 'Pro-Race' cervical cap, 'small check pessary'. It was one of a number of organisations which campaigned successfully in the 1920s for the withdrawal of a Ministry of Health prohibition on giving birth-control advice at Maternity Centres. From 1922–45 its organ was *Birth Control News*. It always was under the unquestioned control of Marie Stopes; in 1930 it co-operated with the **National Birth Control Association**, but broke off links when it became clear that it would not be the exclusive leading organisation. Although the clinic survived its founder, the Society did not as such, though its principles are maintained by Marie Stopes International.

(Leathard, 1980)

Society for Founding and Maintaining Homes for Working Girls in London

Founded in 1878, the aim of the Association was 'to help those who help themselves' by providing homes in various parts of the Metropolis for those who were striving to gain an honest living but were homeless. The Society also ensured that their charges were surrounded by Christian influences and received friendly guidance.

Society for Promoting the Employment of Women

One of the earliest women's organisations, the Society was established in 1859 to promote the training and employment of women. The Society, set up by Jessie Boucherett, Barbara Bodichon and Adelaide Anne Procter, had its headquarters in Langham Place, London, W, where the *English Woman's Journal* was based. Lord Shaftesbury was its first president, and branches were established throughout the country to train and find employment in a range of occupations, though it eschewed the heavy manual ones. Women were trained in book-keeping and clerical work, hairdressing, nursing and house decoration. A law stationers' office was manned by women, and compositors published, amongst others, feminist periodicals and pamphlets. The lack of basic skills in the 3Rs led Boucherett to organise courses, lasting two years, for women who wished to take up clerical work. Essentially a middle-class, politically Conservative movement, the Society laid great emphasis upon self-help, public spiritedness and modest rewards for the employees' efforts. The Society was able to operate only on a comparatively small scale, though its influence was impressive. It considered that the answer to the surplus of women in Britain was to encourage emigration to the Colonies, but this movement was not taken up at the time. By 1875 it had moved to 22 Berners Street, London, W.

(Holcombe, 1973)

Society for Promoting Female Welfare

Started in 1866 under the title of the National Office of Institutions for Women and Girls of Good Character by Mrs Goode, wife of the Dean of Ripon, its object was to promote female welfare by providing knowledge of institutions for women and girls of good character. A register was kept of these institutions, such as homes for incurables, convalescent homes for children, girls' industrial schools and homes for the aged poor. A register of vacancies was centrally kept and the Society informed institutions of applicants hoping for admission. It also later helped to find situations for their inmates on leaving and sold items from the homes at the Office. The title of the National Office was changed to the Society for Promoting Female Welfare in 1872 when men as well as women were admitted to its management. Its headquarters were at 22a Devonshire Street, London, W.

Society for Promoting the Return of Women as Poor Law Guardians

The Society, also known as the Women Guardians Society, was formed in February 1881 to encourage women to become Poor Law Guardians. In 1875, Martha Merrington had been elected for Kensington but by 1881 there were only two Women Guardians. The Society helped to form local committees and gave advice on electoral strategies as well as providing literature and speakers for meetings. Louisa Twining was one of its many active members. The Society was successful in its activities. From 1894, when the Local Government Act removed many of the barriers, including property qualifications, affecting women, the number of Women Guardians increased from 119 to 692. The **Women's Local Government Society** thereafter took over responsibility for the election of Women Guardians and the Society for Promoting the Return of Women as Poor Law Guardians ceased to exist from 1904.

(Hollis, 1987)

Society for Promoting Women as County Councillors *see* **Women's Local Government Society**

Society for the Encouragement of Home Study

Established in 1869, originally for the benefit of young women who had left school but who wished to continue with their studies, it became a Society suitable for young ladies resident in London or in the country. It encouraged regular study by the awarding of annual prizes. The topics were very varied and the essays were examined by university professors, who stipulated the reference books to be used. The Society itself was able to lend members books. One of the highlights of the annual general meeting, which lasted for three or four days in late May or early June, was the opportunity for essayists to meet their tutors. This was also the occasion for visits to places such as Kew Gardens, the National Gallery and the Royal Academy of Arts, as well as events such as the International Health Exhibition and the Handel Festival. Boarding at a very moderate rate was provided by ladies of the organising committee of the Society for young women who were from out of town. Subscriptions for the Society were a guinea per annum and the annual general meeting five shillings.

Society for the Equal Ministry of Men and Women in the Church *see* **Society for the Ministry of Women in the Church**

Society for the Ministry of Women in the Church

This began in the 1940s as the Society for the Equal Ministry of Men and Women in the Church; its name was changed in 1958. It is an interdenominational society. Its original aim was to promote the ordination of women, especially in the Church of England, and it worked closely with the **Anglican Group for the Ordination of Women to the Historic Ministry of the Church** to that end. It has published a newsletter since 1954. Prominent members in the 1950s included Edith Picton-Turbervill, Kathleen Bliss and Miss C. S. Wilkinson. It was still active at the time of writing.

Society for the Oversea Settlement of British Women

Formed in 1919 from the merging of the **British Women's Emigration Association**, the **Colonial Intelligence League** and the **South African Colonisation Society**, this body continued to encourage the emigration of suitably qualified women to the Dominions and other Commonwealth countries, but now as a quasi-governmental organisation, partially funded by a government grant. Its Education Department was particularly successful between the Wars in placing women teachers of specialist subjects. Its activities included obtaining grants for wives and children to accompany emigrating husbands. In 1962 it changed its name to the Women's Migration and Oversea Appointment Society.

(Hammerton, 1979; Monk, 1963)

Society for the Promotion of Co-operative Housekeeping and Household Service

Interest by the **Fabian Women's Group** in schemes of co-operative housekeeping had been inspired by works such as Edward Bellamy's *Looking Backward* (1888) and Marie Howland's *Tomorrow: A Peaceful Path to Social Reform*. Ebenezer Howard, with his colleagues Raymond Unwin and Barry Park, subsequently planned residential units which shared basic domestic services, as was realised in his Garden City at Letchworth in Hertfordshire. The Society for the Promotion of Co-operative Housekeeping and Household Service was founded by the feminist Alice Melvin before the First World War. The Society had two main aims: to extend the principle of co-operation in housekeeping and to widen the scope of housecraft as a profession for educated women. Melvin hoped that associations could be formed to rent adjacent houses where communal rooms would be available for members, or to develop estates such as garden cities where public facilities, such as meals and day nurseries staffed by qualified nurses, were available. However, the plan was limited in scope: it was essentially one designed to serve professional women in middle-class areas such as Hampstead Garden Suburb and proved to be somewhat utopian in character.

(Dyhouse, 1989)

Society for the Promotion of Women's Suffrage

This may be a garbled reference to the **National Society for Women's Suffrage**; in the early 1870s Amelia Edwards and Emma Paterson seem to have been involved. Emilia Dilke mentions it in 1889.

Society for the Provision of Birth Control Clinics

This was the name taken in April 1924 by the Walworth Women's Welfare Centre, which had opened Britain's second birth-control clinic in November 1921 at 153a East Street, Walworth, London, SE17. The Centre had been established largely on the initiative of the Malthusian League, although it severed this connection in September 1922. The new Society encouraged the formation of new clinics and local groups. Its aims were 'to establish and support, in the interests of social welfare and for the relief of poverty, clinics in which instruction in the most satisfactory method of contraception will be given to married women in poor circumstances by registered medical practitioners (preferably women) assisted by qualified nurses'. It depended on voluntary contributions; prominent supporters included Lady Denman, Lord Buckmaster, John Maynard Keynes, Harold Cox, Sir John Sumner and Lord Gerald Wellesley. Its chairman was Margery Spring Rice. The first clinic outside London was in Wolverhampton, and this was closely followed by other clinics mainly in the north of England. Its main specialist adviser was Dr Norman Haire, a practising gynaecologist, who disagreed with Marie Stopes about contraceptive methods, thus incurring the enmity of the **Society for Constructive Birth Control and Racial Progress**. The Society's members were vigorous in campaigning for the acceptance of the principle of birth control, and against the Ministry of Health's prohibition of the giving of birth-control information by Maternity Centres. By 1930 the Society had 11 affiliated clinics. The perceived need for an overall co-ordinating body led it to join with other bodies in the **National Birth Control Association** in 1930; it fully amalgamated with the Association in 1938.

(Leathard, 1980)

Society for the State Registration of Trained Nurses *see* **National Council of Nurses of the United Kingdom**

Society of American Women in London

Founded in 1896 at 5a Pall Mall East, London, the object of the Club was to provide a social and intellectual centre for American women resident in or visiting London. It was affiliated with the General Federation of Women's Clubs in America. There were 150 members and residential accommodation was available. Resident members paid an entrance fee of two guineas with a matching annual subscription, and non-residents paid one guinea entrance fee and a matching annual subscription.

Society of Female Artists *see* **Society of Women Artists**

Society of Lady Artists *see* **Society of Women Artists**

Society of Women *see* **Manchester and Salford Society of Women in the Bookbinding and Printing Trades**

Society of Women Artists

Opportunities for women to obtain fine art education were strictly limited in the first half of the nineteenth century and it was not until 1861 that female students were conditionally accepted by the Royal Academy of Arts in London. The exhibiting and selling of paintings was often confined to membership of the exhibiting body and it was in an attempt to overcome this difficulty that a group of 40 women set up the Society of Female Artists in 1857. The first two exhibitions were organised by Mrs Harriet Grote, wife of the distinguished Greek historian. Another enthusiastic committee member was Mrs Jenny Lind Goldschmidt, the Swedish soprano. In the Society's exhibitions, the majority of paintings were by women but men were not excluded. When the Society was reorganised in 1865, it received support from its patron, HRH the Duchess of Cambridge, and 17 titled ladies.

As access to professional training became possible in the 1860s and 1870s, the Society made an effort to raise standards. It changed its name to the Society of Lady Artists in 1869, when membership was limited to 23, with full membership limited to professional women, though an Associate category was introduced in 1878. Provision was made for giving financial help to sick and needy members. It changed its name again later in 1899 to the Society of Women Artists. In 1916, membership was two guineas and non-members could submit two works for exhibition for a fee of 10s.6d., which was returned if neither was accepted. In the 1930s, the Society consisted of 134 members. It never owned its own gallery or club house. From 1866–73, it exhibited at 9 Conduit Street, London, from 1896 to 1923 at 6a Suffolk Street, Strand, London, and from 1923, at the Royal Institute Galleries. The Society is still in existence at the time of writing.

(Deepwell, 1996; Gaze, 1997)

Society of Women Employed in Bookbinding

It was formed in 1874 by Emma Paterson as a trade union representing women bookbinders, excluded from the traditional men's unions, with whom it was often in disagreement. In 1875, Emma Paterson was succeeded as secretary by Eleanor Whyte, a working bookfolder, who remained in office until shortly before her death in 1913. Membership was never more than 300; entry was a shilling, and subscription was 2d. a week. It held regular meetings and an annual soirée at Foresters' Hall. It appears in practice to have been more of a benefit society, providing death and sickness benefits, than a union campaigning over pay and conditions.

(Tidcombe, 1996)

Society of Women Housing Estate Managers

A national Association of Women Housing Workers, inspired by the work of Octavia Hill, was formed in 1916. It had submitted evidence to the Ministry of Health in 1919 and 1920 on the problems of slum areas on the working of the 1915 Rent Restrictions Act. The growth of municipal housing was followed by the employment of women property managers by local authorities from 1927. These managers formed their own organisation in the following year, and in 1932 it became the Society of Women Housing Estate Managers. A certificate especially to cater for their needs was offered by the Chartered Surveyors' Institute from 1933. In 1937 it changed its name to the Society of Women Housing Managers. Men working as housing managers in local authorities had founded their own association, the Institute of Housing, but in 1948 the Society of Women Housing Managers omitted the word 'Women' from its title and both men and women were henceforth eligible for training and membership. The two associations merged in 1965 and became the Institute of Housing Managers.

(Brion, 1995; Dyhouse, 1989)

Society of Women Journalists

By the last decade of the nineteenth century there were opportunities, particularly for middle-class women, to follow the career of journalism. Prominent among them were correspondents such as Flora Shaw of *The Times* who reported from South Africa and Emily Crawford of the *Daily News* in Paris. Women's magazines had increased the space devoted to women's topics and this presented further opportunities. In 1894, Joseph Snell Ward, editor of the *Gentlewoman*, formed the Society of Women Journalists, initially in an upstairs room of the London Shoe Company, Bond Street; its headquarters later was at Granville House, Arundel Street, Strand. His aim was to unite women journalists for mutual protection and advancement and to encourage literary achievement. There were three classes of members: those who had worked in a department of a recognised journal or magazine as a paid contributor for at least two years, probationers who had worked as above for one year and honorary members, who took an interest in upholding a high standard in journalism. Subscriptions were one guinea a year and a benevolent fund was available for needy members. It had a Council of 25 women members and mounted a programme of debates, lectures and receptions. The Society was in a flourishing state at the beginning of the twentieth century. In 1952 its title was changed to the Society of Women Writers and Journalists. The house magazine, begun in 1910, was entitled the *Woman Journalist* and was published three times a year. In 2000, its title was changed to the *Woman Writer*.

(Rubinstein, 1986)

Society of Women Musicians

This Society, formed in 1911, reflected the growing number of women composers and executants. One of the main grievances of composers especially was the difficulty in hearing their works performed in public.

The Society's main objective, therefore, was to 'raise the art of music by breaking through existing conventions', provide professional women with a suitable centre for the discussion and criticism of musical matters, and to act as forum in bringing together composers and executants. Its first president was Liza Lehmann, the composer of the song cycle, *In a Persian Garden*.

(Lehmann, 1919)

Society of Women Writers and Journalists *see* **Society of Women Journalists**

Society of Women Welders

It was formed during the First World War on an initiative of the Women's Service Bureau of the **London Society for Women's Suffrage**, to take advantage of the wartime skill shortages. It ran a training school for welders, and had its own badge, representing a welding spark. However, by the 1920s, the Society was faced with unemployment for a number of its members. It complained of skilled workers being forced to accept unskilled jobs. Welfare officers recommended that many of them should return to domestic service. It does not appear to have survived to the end of the decade.

(Lewis, 1984)

Soho Club

This Club, at 59 Greek Street, London, which was in existence in the late nineteenth century, was for working-class girls. Classes were held in a range of subjects including English literature and history, dances were held every Friday and concerts and dramatic productions were arranged. Visits to the country were made usually on Bank Holidays and friends of the Club opened their country houses to the girls, sometimes offering short stays there. In 1893 there were 200 members, representing a wide range of occupations.

Somerville Club

The Club was established in 1878 at 231 Oxford Street, London, as 'a meeting place for solitary workers.' Both entrance and subscription fees were fixed at the moderate rate of five shillings. It had a serious purpose, holding regular lectures and discussions. There was an ample reading room with newspapers and periodicals as well as a free lending library. It was obliged to close in 1887 but it was re-formed in the following year as the **New Somerville Club**.

(Crawford, 1999)

Soroptimist Clubs

'Soroptimism' (Latin *soror* = sister plus 'optimism') first started in the USA in 1921, and was built on the principle of a local grouping of the most outstanding women in each business and profession in a given district, along the lines of the Rotary Clubs (at that time strictly male). The first British club was the Greater London Club, founded in 1923. Its aim was to advance the status of women, maintain high ethical standards in business and professional life and develop unity amongst Soroptimists in all countries. Soroptimist Clubs consist of one member from each business or profession who is over 25 years of age and of high standing in her field of work. Activities include providing housing for retired business or professional women, participating in the work of clubs, particularly with handicapped children, and co-operating with other women's organisations. The organisation is concerned with issues such as human rights, the environment, health, the status of women, economic and social development and the fostering of international goodwill and understanding. In 1930 the Soroptimists united with another group of clubs with the same aims, the Venture Clubs of Great Britain, but continued to be known simply as Soroptimists. Their publication is the *Soroptimist* which is issued eight times a year.

South African Colonisation Society *see* **British Women's Emigration Association**

Spiritual Militancy League

This appears to have been in existence from about 1913 to 1916, with the general aim of campaigning for women's suffrage from a religious standpoint.

Stamp Section (GPO) Lady Assistants' Association

The Association was founded in 1920 for staff taken over by the Post Office Stores department in 1914. Initially it had 17 members. The Postmaster-General, Albert Illingworth, refused permission for the Association to affiliate to the Council of Post Office Clerical Association as he considered that their work was not of a clerical nature. The Association ceased to exist in 1922 when it joined the Civil Service Clerical Association.

Standing Conference of Women's Organisations

Founded in 1942, the Standing Conference's objectives were to promote a common meeting ground for the exchange of information on matters of local or national interest to all women and family organisations, to promote interest in civic and social matters affecting policy, and link with councils for voluntary service and to promote women's interests. Representatives serve on many regional, national and European bodies which promote the interests of women and family organisations.

Standing Joint Committee of Industrial Women's Organisations *see* **National Joint Committee of Working Women's Organisations**

Stansfeld Trust

This Trust was set up to commemorate the work of Sir James Stansfeld (1820–98), Radical politician and women's champion. Its aim was to promote the equality of men and women before the law, and to make suggestions for altering existing law to achieve this end. The Trust consisted of monies subscribed to further this work.

Societies and private individuals subscribed a guinea a year and in return received reports on Bills before Parliament which affected the interests of women. A member of the Trust read all Bills as they were published and her reports on points which concerned women were immediately printed and posted to subscribers. In 1901, the honorary secretary and treasurer was Miss M. J. Shean, 15 Phillimore Gardens, Kensington, and the scrutineer of parliamentary Bills was Miss Chamney, 3 Chenies Street Chambers, London, WC.

Status of Women Committee

In 1935 the League of Nations appointed a Committee of Experts to study the civil and political status of women worldwide (their economic position was excluded as falling within the remit of the International Labour Office), and responses were solicited not only from governments but also from international women's organisations. In Britain the Status of Women Committee was formed on the initiative of the **National Council of Women**, to co-ordinate responses and forward them through the International Council of Women. From the outset, the distinction between economic and other rights was sharply contested, but given the League's decision, it had to be accepted. During the Second World War the work of the Committee virtually ceased, but was revived in 1945 less as a mechanism for reporting to the newly-formed United Nations Organisation than as a national body co-ordinating the work of organisations campaigning for women's rights. Member organisations included the **Association for Moral and Social Hygiene**, **British Commonwealth League**, **Society for the Ministry of Women**, **Married Women's Association**, **National Union of Women Teachers**, **National Women Citizen's Association**, **Open Door Council**, **St Joan's Alliance**,

Six Point Group, **Society of Women Journalists**, **Suffragette Fellowship**, Women for Westminster (campaigning to return women MPs), **Women's Freedom League** and the British Section of the Women's International League for Peace and Freedom. Initially the **National Association of Women Civil Servants**, the **National Council of Women** and the **National Federation of Business and Professional Women's Clubs** had observer status. The first post-War chairman was Thelma Cazalet-Keir, whose successors included Vera Laughton Mathews and Joan Vickers. The Committee prepared manifestos for all general elections from 1950–66 urging the consideration of such issues as equal pay and employment rights, equal political and civil rights, an equal moral standard, equality in taxation and national insurance, equal rights in the marital home and equal ministry within the Church. The Committee ceased activity in 1980.

Student Nurses' Association

This Association was founded in 1925. Members consisted of probationers working in hospitals. In 1969, the Student Nurses Association became the Student Section of the **Royal College of Nursing**.

Suffrage Atelier

This was founded in 1909 by Laurence and Clemence Housman as an 'Arts and Crafts' society to support women's suffrage. It mainly provided material for the militant suffrage societies, especially the **Women's Freedom League** and the **Women's Social and Political Union**. Active members included Catherine Courtauld, Edith Craig and Hope Joseph. Its productions in the form of cartoons, posters and the like, often show a striking expressionist modernism unusual in political propaganda at this time. Its small but very individual logo was an Attic owl. Its activity seems to have ceased after August 1914.

(Crawford, 1999; Tickner, 1987)

Suffrage Club

Established at 3 York Street, London, for men and women but with residential accommodation for women only. It closed in 1916 or 1917.

Suffragette Fellowship

Founded as the Suffrage Club by Edith How-Martyn in 1926, this was partly an old girls' association of militant suffragists (mainly from the **Women's Social and Political Union**, despite its founder's background as a leading member of the **Women's Freedom League**) and partly a body formed to preserve records and memorabilia of the militant suffrage movement. In 1951 the Fellowship's collection was deposited with the

London Museum (now the Museum of London). Its activity seems to have effectively ceased in the late 1970s with the death of Enid Goulden-Bach, its last organiser.
(Mayhall, 1995)

Suffragettes of the Women's Social and Political Union *see* **Women's Social and Political Union**

Suffragists' Vigilance League

The only mention of this organisation appears to be in connection with the one-off periodical the *Suffragist*, published in 1909.

Territorial Forces Nursing Service

In 1908, R. B. Haldane, the Secretary of State for War, set up a Territorial Force to supplement the regular Army. Elizabeth Haldane, his sister and vice-president of the **Society for the State Registration of Nurses** and member of the Army Nursing Board, envisaged a reserve of nurses who would help in military general hospitals in a future war. The nurses would serve in any one of 23 hospitals located in the main cities. Some 3,000 recruits were needed. Nurses, who had to show proof of three years' training, were between the ages of 23 and 50. Peace-time work would be on a voluntary basis but regular payments would be made for war service. Miss Browne was appointed Matron-in-Chief. Soon after the First World War broke out, the members of the TFNS were dealing with wounded English and Belgian soldiers and by 1915 it was necessary to recruit a further 3,000 nurses. Some 400 saw service in France and Belgium.
(McGann, 1992)

Theatrical Ladies' Guild

Founded in 1891 at the home of Mrs Keeley a former actress, in order to offer assistance to destitute members of the profession who were about to become mothers. The president was Miss Fanny Brough who was 90 years of age, and Henry Irving was a patron. Sewing classes were held weekly in London and in the provinces. In 1893 there were 300 members. Its address was 3 Bayley Street, Bedford Square, London, WC.

Three Arts Club

Situated at 19a Marylebone Road, London, W, this Club was for women in the professions of painting, music or drama. The entrance fee was one guinea, with subscriptions of a guinea for town members, and half a guinea for country members.

Time and Talents Guild

Its objective was 'to seek through fellowship, prayer and service to bring the Spirit of Christ into every part of life'. It was formed in 1888 on an initiative of Minna Gollock, Mrs Abel Smith and Mrs J. H. Tritton, under the auspices of the **Young Women's Christian Association**. It was originally a philanthropic organisation, organising girls into devoting time to helping the poor by making dolls, preparing scripture texts in different languages for foreign missions, etc. It was organised in local centres, the first of which was started in Edinburgh in 1889. By 1891 there were 22 centres, with a membership of over 1,100. From 1898, it ran a Settlement at 187 Bermondsey Street, London, SE1. It was still active in the 1980s.

(Daunt, 1989)

Toc H Women's Association

The Rev. Philip (Tubby) Clayton, an army chaplain serving on the Western Front in the First World War, established a rest house in 1915 for troops at Poperinghe, Belgium – Talbot House, known as Toc H, in memory of his brother who had been killed in action. After the war, Clayton established a Talbot House in London and began the movement which was to be a living demonstration of fellowship and self-sacrifice, based on Christian fellowship. Members used their skills in all forms of personal service and gave help where needed to other organisations. The movement quickly spread with several hundred branches being formed at home and overseas. A Women's Association with identical aims to those of Toc H was established in 1922. By the 1960s there were 550 branches in the United Kingdom served by seven regional secretaries, with 160 branches overseas. It promotes friendship and respect for human rights.

Town and Country Club

This Club, founded in 1897 at 34 Berners Street, London was open for gentlewomen with short-term accommodation provided. The subscription fee was 10s 6d. There were 650 members in 1901.

Townswomen's Guilds

These were the brainchild of Eva Hubback and Margery Corbett Ashby, and were intended to be the urban equivalent of rural Women's Institutes, giving women in towns the opportunity to get together, compare notes, gather information, take civic action and generally mix socially. The first Guild was started at Haywards Heath, Sussex, in January 1929; at first it was called the Women's Town Institute, but the name was soon changed after the **National Federation of Women's Institutes** expressed concern at possible confusion. Other Guilds soon followed, and they were linked in a National Federation. Members are involved in many activities including consumer work, energy conservation, teaching communication skills, and there is a wide range of lectures in the arts, social studies, current affairs and sports. The Guilds' organ is the *Townswoman*. In 1999 the Guilds claimed a national membership of 80,000.

(Merz, 1988)

Travellers' Aid Society

This was started on the initiative of Mr Hind Smith in October 1885 at No.19 Exeter Hall, London, with the purpose of drawing up a scheme for aiding female passengers and to report to a future meeting of representatives of existing societies. Those present at the meeting included Mrs L. Ormiston Chant, Mrs Percy Bunting, Mr Ralph Thicknesse and Mr W. A. Coote. The meeting appointed a committee chaired by Mrs Maddison; the 'existing societies' included the National Vigilance Association, the **Girls' Friendly Society**, the **Metropolitan Association for Befriending Young Servants**, the Reformatory and Refuge Union and the **Young Women's Christian Association**. The bulk of the work involved fell to the last named, and it was at a YWCA office at 16a Old Cavendish Street, London, that the executive committee first met. The Society established offices at the major London railway termini with the aim of helping young women new to the city to settle, and to keep them from exploiters, mainly sexual. As one contemporary wrote, 'There is a certain class of girls who lose their heads the moment they enter a railway station.' The work of the Society was not confined only to helping Christian women. For example, it collaborated closely with the **Jewish Association for the Protection of Girls and Women** in meeting ships at ports and ensuring the newly arrived girls' safety. After the First World War, funding declined considerably and the Society merged with the National Vigilance Association in 1939; however, the Association continued to carry on its work under its name until the Association itself suffered a financial downturn in 1952.

(Burdett-Coutts, 1893)

Twentieth Century Club

This Club, founded at 24–29 Stanley Gardens, London in 1902, provided suitable accommodation for educated women workers, especially those engaged in professional, educational, literary, secretarial or other similar work. The Club had a dining room, sitting room, a library and residential accommodation. It was a profitmaking concern with three male directors of the company. It no longer existed in 1924 and was wound up in 1937.

Ulster Society of Women Artists

Established in Belfast in 1957 to promote and encourage a high standard of art in Ulster. Potential members had to submit four of five works to the Society's Selection Committee. If elected, the candidate was given a mark of distinction. In 1997 the Society had 100 members.

Union of Catholic Mothers

Stemming from the **Catholic Women's League** in 1913, the Union was based on the ideas of the **Mothers' Union** which provided a forum for praying and working together, and upholding the Christian ideals of marriage and family life. The Union is based on parishes but there is a national organisation which keeps in contact with parliamentary and international issues which may affect its members. Study days are organised and welfare officers give assistance to families in trouble. The Union issues a magazine, *Catholic Mother*, twice a year.

Union of Jewish Women

The Union was formed in May 1902 after the first national Conference of Jewish Women in London. Its aim was to promote the social, civil, moral and spiritual welfare of Jewish women, and to encourage practical co-operation and co-ordination between Jewish workers in London and the provinces. The Union also hoped to disseminate sound principles of action in connection with philanthropic work, considering that any aid given should lead to self-respect and self-support. It did not wish to undertake any activities which competed with existing societies or charitable institutions but rather sought to provide, through its information bureau, a means of bringing existing societies into touch through arranging meetings and conferences from time to time in different towns. The Union also administered a loan fund for training educated ladies for various occupations. Posts were found for a growing number of middle-class Jewish women, such as teachers, governesses and clerks. Branches and correspondents of the Union of Jewish Women were established in nearly all the provincial towns, European capitals, the United States and British Colonies. The membership in 1916 was 1,600. The president at that time was Mrs Nathaniel L. Cohen and the treasurer Mrs A. Q. Henriques.

(Black, 1988)

Union of Practical Suffragists

This was founded in 1896 to act as a ginger-group within the **Women's Liberal Federation**, which at the time was perceived as being insufficiently vigorous in its pursuit of the parliamentary vote for women. It took a firm stance on a number of feminist issues, but the core of its campaign was the vote, on which it published a number of leaflets written by its prominent members, who included Clementia Taylor (Mrs P. A. Taylor), Mary Priestman, Florence Balgarnie, Eva McLaren, Harriot Stanton Blatch, Ursula Bright and Dora Montefiore. Its main stipulation was that Liberal candidates should only be supported if they made a commitment to women's suffrage, and when the **Women's Liberal Federation** agreed to this in 1903, the Union dissolved itself.

(Crawford, 1999; Leeds, 1898)

Union of Private Governesses

By the beginning of the twentieth century, the majority of governesses were middle-class women with no academic qualifications. Their salaries lagged behind those of other women teachers and tended to be lower than those of better qualified private governesses. In 1912, a group of private governesses met at Queen's College, Harley Street, London to organise a Union of Private Governesses to safeguard their interests.

Union of Women Matchmakers

After the publication on 23 June 1888 of an article in the *Link* written by Annie Besant describing the pay and conditions of women matchmakers in Bryant and May's East End factory, three of them were dismissed by the management for allegedly talking to Besant. When further sackings followed a refusal by some employees to publicly refute Besant's article, the 1,400 women employed at the factory went on strike. Attempts by the employers to break the strike were thwarted by the moral and financial support received by the women, including essentially men's organisations such as the London Trades Council. By the following month, the firm had conceded defeat, even going so far as to 'hope the girls would form a union'. On 27 July 1888, the first meeting of the Union of Women Matchmakers was held. Annie Besant was its secretary and the Union had its own headquarters. By October, there were 666 members but victimisation of the officers of the new union by the owners led to changes. By the end of the year it had changed its name to the Matchmakers' Union and membership was open to both men and women. It continued in existence until 1903.

United British Women's Emigration Association

This was originally formed in 1884 as the United Englishwomen's Emigration Register (later Association) with the intention of continuing the work of the **Women's Emigration Society**, originally sponsored by Louisa M. Hubbard, who however soon abandoned it. From 1885, under its new name, it was effectively organised by Adelaide Ross and the Hon.

Mrs Ellen Joyce, and later, Grace Lefroy. Mrs Joyce previously had been instrumental in helping families to emigrate through the Winchester Emigration Society and had organised an emigration department for the **Girls' Friendly Society**. It rapidly became the main women's emigration society, taking over the **Colonial Emigration Society** in 1892. In 1901 it became the **British Women's Emigration Association**.

(Hammerton, 1979; Monk, 1963)

United Englishwomen's Emigration Association [Register] *see* **United British Women's Emigration Association**

United Free Church of Scotland Women's Foreign Mission

For the recruitment of women missionaries who were, ideally, for tropical countries, about the age of 25. Different requirements for different areas were described. For Zenana work in India, a teaching qualification and a good education were desirable. For work in Africa, industrial training was sometimes necessary. In theory, missionaries were supposed to offer themselves for life, but in practice only six months' notice was required to terminate an appointment. Leave in Britain varied according to the climatic conditions of the country to which a missionary was posted. The type of work undertaken was in boarding schools, Zenana missions, day schools, medical missions and rural development. Missionaries were under the supervision and control of the Mission Council of the region.

United Kingdom Federation of Business and Professional Women *see* **National Federation of Business and Professional Women's Clubs**

United Sisters' Benevolent Society

In 1885, the Rev. J. Frome Wilkinson started a United Sisters' Friendly Society for working women on the principles of the Oddfellow and other societies already existing for men. The Society was national, and, consisted of a series of local Courts which controlled their local affairs; The local Courts, which were self-supporting, sent representatives to a policy-making central governing body known as the Annual Committee. In addition, each Court had its own Committee. By means of a small payment, members could secure a weekly allowance in times of sickness. As all the Courts were affiliated, should a financial strain be put on an individual Court, it could be assisted by the others. Louisa M. Hubbard, a champion of poor gentlewomen, established a Work and Leisure Court of the Society, mainly for working professional women. Members paid a higher premium and received increased allowances. It admitted only single women and widows, but married women were eligible for the other Courts. When Wilkinson resigned as president of the Society in 1895, he was succeeded by a woman and the organisation became henceforth almost entirely managed by women.

(Pratt, 1898)

United Suffragists

This group was founded on 6 February 1914 by Henry Woodd Nevinson, Evelyn Sharp and Gerald Gould. Its members were mainly militant suffragists, usually disenchanted former members of the **Women's Social and Political Union**, including Hertha Ayrton, Edith and Israel Zangwill, Daisy and George Lansbury, Louisa Garrett Anderson and Laurence Housman. They were soon joined by the Pethick-Lawrences, and in August *Votes for Women* became the official organ of the United Suffragists. Its militancy was of a moderate brand, and it was in general more friendly to the organised labour movement than was the Women's Social and Political Union; it collaborated not only with trade union leaders but with Sylvia Pankhurst and her East London Federation. It soon had a number of branches throughout Britain. It continued campaigning for women's suffrage and other women's issues after the outbreak of war, and in 1916 it joined the **National Union of Women's Suffrage Societies** and **Women's Freedom League** in lobbying for the vote. It ceased campaigning on the passing of the Representation of the People Act in February 1918.

(Joannou and Purvis, 1998)

United Women Bookbinders' Union

Formed in 1892 under the leadership of the **Women's Trade Union League**, the Union organised semi-skilled and unskilled labour in printing and paper trades. It gained the support of the Printers' and Stationers' Warehousemen, Cutters and Assistants' Union and became a separate branch of it in 1902.

(Bundock, 1959)

United Women Bookfolders' Union

Founded in about 1892, this had approximately 500 members. Its secretary was Mrs E. Bond, and it had offices at 28 Poppins Court, Fleet Street, London. It does not appear to have been a long-lived organisation.

(Tidcombe, 1996)

United Women's Insurance Society

This women's approved Society, situated at Murray House, Vandon Street,

Buckingham Gate, London, SW1, offered services to 'household workers, nurses, matrons, secretaries, etc'. It was in existence in 1934.

University Women's Club

Established in 1887 as the University Club for Ladies at 32 George Street, Hanover Square, London, its members consisted of university graduates, medical practitioners of the United Kingdom and certain other diploma holders. Women who had distinguished themselves in promoting education, science or the arts were eligible to join as special members. By 1904, the Club had 450 members. It changed its name to the University Women's Club in 1921 and had moved to 2 Audley Square, London, W1 by 1922.

Victoria Club (1)

Founded in 1894 at 145 Victoria Street, London, to provide a town house for country ladies. As a purely social Club it was stated that members should not express political, educational or 'advanced' views. It claimed to be the first of its kind. When it moved to 3 Old Cavendish Street, its facilities included a drawing room, a reading room and a dining room. The annual subscription was two guineas. Rooms were available for members to interview servants and dressmakers.

Victoria Club (2)

Founded in 1876 at 25 Regent Street, London, it offered refreshments and waiting rooms for its members. The entrance fee was one guinea and membership two guineas per annum. The prospectus stated that 'Ladies wishing to dress for an evening engagement can have the assistance of a maid without extra charge, except for hairdressing.' It is unclear whether or not this is a predecessor of the previous entry, or an entirely different organisation.

Victoria League

Founded in 1901 at 2 Millbank, Westminster and dedicated to the memory of Queen Victoria, the League described itself as 'an association of women of the British Isles who are in sympathy with Imperial objects and desire a close union between the different parts of the Empire'. It claimed to be non-political, although the inaugural meeting was held at 10 Downing Street at Arthur Balfour's invitation. The prime movers were Violet Cecil and Edith Lyttelton, both of whom had visited South Africa at the end of the nineteenth century. From its inception, the League was an exclusive coterie of wives of high-ranking politicians of both the leading parties. Edith Lyttelton became the League's first secretary and the funds for the League came from wealthy patrons. It was a centre for receiving and distributing information on different British Dominions which was of particular importance to women. It also gave practical help, such as providing aid to British refugees and Dutch women and children in camps in South Africa.

Professional lecturers were employed to address audiences of industrial workers, agricultural labourers, schools and more general audiences. A lending library was installed and books and newspapers were despatched to adults overseas and children were also catered for. Council meetings were held annually and the Executive Committee remained largely unchanged from year to year. Sub-committees such as those for literature, education and hospitality were set up. By 1914, there was a number of branches in towns and cities in Britain, and there were about 6,000 members. Branches recruited junior members aged 10 or over and it was reported in 1908 that the Harrow Victoria League branch had no fewer that 700 young associates; it became a model for others to follow. The League still exists under the title of the League of Commonwealth Fellowship.

(Bush, 2000)

Victoria Press

This was established by Emily Faithfull in March 1860 at 9 Coram Street, London, WC1 (it later moved to Farringdon Street), originally as a printing establishment in which women were to be employed exclusively as compositors, something which the male print unions vehemently opposed then and for at least a century thereafter. It depended largely on the sponsorship of the **Society for Promoting the Employment of Women** and the National Association for the Promotion of Social Science, although it did win support in higher quarters: Queen Victoria appointed Emily Faithfull 'printer and publisher in ordinary'. In fact it had begun publishing in a small way very soon after its formation; its first publication in 1861 was the *Victoria Regina*, edited by Adelaide Procter. One of the Press's most important products was the *Victoria Magazine* (1863–80), which in its time was widely read and highly influential: it is the only British feminist magazine to be included by the American William Poole in his famous Index. In 1867 Emily Faithfull left the Press, and from then until 1870 its proprietor was William Wilfred Head. It appears to have ceased operating some time in the mid-1880s.

(Fredeman, 1973)

Victorian Club

Founded in December 1897 at 4 Hope Street, Edinburgh, it was available for women who were engaged or interested in preparing for the various professions. Other ladies could be elected at the discretion of the Committee. The Club's aim was to provide a convenient meeting place, to promote social intercourse and interest in women's work generally. Residential accommodation was available for short visits. Management was by a committee of members. Fees were modest. The entrance fee for town members was five shillings and country members two shillings and sixpence. The annual subscription for town members was 15 shillings and country members 7s 6d. The Club was managed by a committee of members. The president in 1906 was Miss F. H. Simpson and the honorary secretary Miss S. B. Balfour.

Voluntary Aid Detachment

In 1908, the Territorial Army established a supplementary military medical organisation and in the following year the War Office issued a 'Scheme for the Organisation of Voluntary Aid in England and Wales' for men and women. The VAD, which functioned on a county rather a national basis, worked with the British Red Cross Society and later, the Order of St John of Jerusalem also. The women's detachments served in rest stations and helped on wards and were trained in first aid and home nursing. A detachment was sent to the Balkans in 1912 . By 1914, there were 46,000 women members. A number served in France and Belgium and in the following year acted as nurses, under supervision. By 1918, they were serving in all theatres of war.

When the Second World War broke out in 1939, members of the VAD were posted to naval, military and air force hospitals. An attempt by the Army Council in May 1942 to assimilate the VAD into the **Auxiliary Territorial Service** met with fierce opposition and the plan was dropped. The first contingent of members were drafted overseas in 1944 and served in north-west Europe, the Mediterranean and other theatres of war. By 1946, there were more than 8,000 members of the VAD.

(Bowser, 1917)

Votes for Women Fellowship *see* Women's Social and Political Union

Wantage Club

Begun by the Rev. John Fenwick Kitto, vicar of St Martin-in-the-Fields, London, in 1897, it was a Club for waitresses and others engaged by hotels, restaurants and such establishments. It was a place for rest and recreation when off duty. Instruction and entertainments were provided in the winter months and there was a library from which books could be borrowed.

West Central Jewish Girls' Club

Established in 1897, with premises at 8a Dean Street, Soho, W, the club was intended to provide recreation in the middle of the West End 'with a view to discouraging girls from harmful amusements'. It organised evening continuation classes offering dressmaking, millinery, reading, writing, basket-making and Hebrew for young working women who lived in the district. Amusements included dancing, debates and concerts. Many ladies, notably Lily Montagu, participated in the work of the club as it was considered that members would benefit from social intercourse between women of different educational backgrounds and occupations.

(Braine, 1902)

White Ribbon Association

Established in 1876 to benefit women and children and to educate them particularly in the dangers of alcohol, tobacco and to protect the family against harm to children. The Association issued a quarterly publication, the *White Ribbon Magazine*.

Whittington Club and Metropolitan Athenaeum

This was launched in 1846 as a mixed Club for men and women, with Douglas Jerrold as president and William Shaen as chairman. Council members included Matilda Hays, Eliza Meteyard and Clementia Taylor.

(Gleadle, 1995)

Woman's Elevation League

This League was active about 1852. Its aim was to promote the social, moral, professional, pecuniary and political elevation of women.

(Gleadle, 1995)

Woman's League for Mothers and Women in Positions of Responsibility

The Woman's League was an association of women to help each other in upholding the high standards of domestic and social life implicit in the Christian religion. To this end, the League urged the education of women, in order that they become aware of their responsibilities and carried out such work in dioceses. The diocesan council was restricted to communicants of the Church of England, but no such restriction was made for the admission of members. By the end of the first decade of the twentieth century, the League had 175 branches.

Woman's Mission to Women

Although the Mission, which was based at 32 Charing Cross Road, London, SW, had a male honorary secretary, its work was undertaken entirely by women. They operated in London among prostitutes, distributing tracts and providing temporary lodgings 'for those desirous of Reforming'. Up to 1885, nearly 7,000 'fallen women' had been placed in homes, 6,845 were found occupations, 1,909 were 'restored to friends', 78 emigrated, 79 were assisted to marry and 5,837 were 'dealt with in other ways'.

Women Cigar Makers' Protective Union

Established in London in 1889 with 750 members, the Union became a branch of the Cigar Makers' Mutual Association in 1895. By 1914, its members had achieved uniform rates of pay for men and women.

Women Guardians Society *see* Society for Promoting the Return of Women as Poor Law Guardians

Women Inspectors

Poor Law

Louisa Twining, an active worker in the field of Poor Law reform, argued the case for the appointment of women Poor Law Inspectors in order to supervise women and children and be concerned with health and domestic arrangements. James Stansfeld, President of the Local Government Board, appointed Mrs Jane Elizabeth Senior, daughter-in-law of the economist Nassau Senior, as the first woman Inspector in 1873. In the following year, she produced a report on the physical, moral and domestic training of girls in schools. Her presence was bitterly resented by the male inspectors; she resigned in 1874 and died three years later. The post was left vacant until 1885 when Miss M. H. Mason was appointed as a Boarding-Out Inspector. Two more were appointed in 1898 and 1901.

Twining also demonstrated the need for women inspectors of sick wards and directed her efforts towards improving the Workhouse Infirmary Nursing Association. Miss Stansfeld was appointed as Poor Law Inspector in the Metropolitan district. In 1910, four nurses were recruited to inspect Poor Law children's hospitals, sick wards of infirmaries and certified homes. Stansfeld became Superintendent of Women Inspectors in 1910 when Miss Mason retired. When the Ministry of Health superseded the Local Government Board in 1919, the inspectors were transferred to the new Department, assisting Medical Officers of Health in inspecting local authorities' facilities. The National Assistance Act of 1962 led to a complete reorganisation of the system.

Prisons and Reformatories

Although the first male prison inspector was appointed in 1857, it was not until 1904 that a lady sub-inspector, Mrs H. E. Harrison, was given responsibility for inspecting girls' reformatory and industrial schools. The Departmental Committee on Reformatory and Industrial Schools (1913) recommended that women should visit boys' as well as girls' schools.

A Departmental Committee on Prisons in 1894 advised on the necessity for appointing a women inspector of prisons. The Home Office acted on this recommendation 14 years later in 1908.

Factories

Two years after the Royal Commission on Labour (1891) had highlighted the conditions of women engaged in industrial work, the Home Secretary, Herbert Asquith, promised the appointment of women inspectors. Miss Mary Paterson and Miss May Abraham were appointed the same year, enjoying equality of responsibilities and duties with their male colleagues. During their first year, Abraham travelled over 3,600 miles and Paterson 4,200, visiting millinery, dressmaking and tailoring establishments. This resulted in the issue of a critical report directed at employers. By 1914, the number of inspectors had risen to 21. At first women inspectors were organised on a district basis, but after 1918 senior posts were created on a regional one. In the 1920s, a single system of administration was adopted for both men and women. When the Children's Branch of the Home Office was formed in 1927 for the inspection of reformatory and industrial schools and children's homes, more posts for women were established.

(McFeely, 1988)

Education

After the setting up of School Boards following the Education Act of 1870, many appointed women as inspectors. In London, a Kindergarten (Froebel) 'Instructor' was appointed and by 1900 the School Board for London alone had 17 women inspectors. The first appointment made by the Committee of the Privy Council on Education was Miss Emily Jones as Directress of Needlework in 1883, and an Inspectress of Cookery and Laundry Work, Miss Mary Harrison, seven years later. The first non-domestic posts were the two sub-inspectors appointed in 1896 to help men HMI in their visits to elementary schools. In 1905, Robert Morant, the Permanent Secretary of the Board of Education, recommended that the six women inspectors should form the basis of a separate Women's Inspectorate and entitled HMI (Women) with a Chief Women Inspector at its head. The Hon. Maude Lawrence, daughter of a former Governor-General of India, was appointed. Seven more women were recruited (they were nicknamed 'the washtub women' by the men) and for the first time two secondary school inspectors. By 1912, there was an establishment of 46 which covered besides elementary and secondary education, technical, further education and training colleges. A third of the women HMI were graduates.

An important reform was introduced in 1919 by the President of the Board, H. A. L. Fisher, when seven women were promoted to the rank of Staff Inspector, each attached to one of the branches of the Inspectorate. Fisher also looked to a considerable expansion in their numbers and equality of opportunity. From 1919, men and women participated in the inspection of secondary schools. Nevertheless, there was still a number of outstanding issues concerning women inspectors. In the elementary schools branch, the District Inspector was invariably a man, and women were only rarely consulted on questions of educational policy. In contrast, women in the Secondary Branch were given a large measure of responsibility and were in charge of full inspections of schools. Women were also underrepresented on inspectorial committees and their salaries were uniformly lower at each grade than men.

Separate recruitment to the Women Inspectorate ceased from January 1934 when aggregation into one service, with higher posts open to both sexes, was commenced. In 1945, the first woman Chief Inspector, Miss F. M. Tann, with responsibility for primary education, was appointed. By the late 1960s, women constituted over a quarter of the English and Welsh Inspectorates, some 142 in number. Equal pay was achieved in 1961 and the first woman Senior Chief Inspector, Miss Sheila Browne, was appointed in 1974.

(Lawton and Gordon, 1987)

Women Police Service

This Service began shortly after the outbreak of war in August 1914 on the initiative of Nina Boyle, who placed an advertisement in the **Women's Freedom League**'s organ the *Vote* appealing for 'Women Police Volunteers', chiefly to help women and children in wartime circumstances. At the same time Margaret Damer Dawson started recruiting 'women police', but when she and Nina Boyle learnt of each other's activities, they decided to join forces, with Margaret Damer Dawson as chief, and Nina Boyle as her deputy. They were joined by Mary Allen, a militant suffragette, in organising women volunteers, training and patrolling in uniform with the grudging permission of the Metropolitan Commissioner of Police. However, divisions soon appeared when the volunteers were invited to help police apprehend prostitutes, when Nina Boyle, Edith Watson and Eva Christy declined to participate.

Damer Dawson and Allen, however, insisted that they should follow orders, however distasteful, and carried the majority of the women with them. They were henceforward called the Women Police Service, and although not sworn in as constables, were granted powers of arrest in matters concerning women.

Also in 1914, the **National Union of Women Workers**, together with members of the **Mothers' Union**, the **Young Women's Christian Association**, the **Girls' Friendly Society** and the Church Army, had initiated their own Women's Patrol Committee under the leadership of Mrs Louisa Creighton, and began patrolling in London and other cities. However, at first they kept their distance from the Women Police Service, many of whose members had been militant suffragettes, and even imprisoned. They came to be known as the Special Patrols, and from 1917 were under the leadership of Sofia Stanley. In 1918 they effectively merged with the Women Police Service, with Mrs Stanley as their Superintendent, to form the first group of officially recognised and paid women police in London, under the control of the Commissioner of Metropolitan Police. However, after the war, there were attempts to disband the women police, which were resisted. The Baird Committee reviewed the situation in 1920 and confirmed the need for policewomen, although the Geddes Axe in 1921–22 was a blow to these recommendations. In June 1922, the Home Secretary announced that 20 Metropolitan policewomen would form the nucleus of a force which would be built up in time. One woman inspector was appointed to cover half of the 700 square miles of the Metropolitan Police District with an assistant, without police rank, to cover the other half. Outside London, by 1933, 37 provincial forces employed women, 15 towns had given up employing them. Some advances were made; the first senior women officer, Miss Dorothea Peto, was appointed to Scotland Yard as Staff Officer to the Commissioner in 1930, and in 1933 three Women Detective Officers were appointed to the Metropolitan Police. In August 1939, 12 days before the outbreak of the Second World War, the formation of the Women's Auxiliary Police Corps was announced. Although recruits were dressed in police uniform, they had no police powers and received no training. By 1943, some 1,204 women had joined, with only 370 regular policewomen in post throughout England and Wales. In 1945, a Woman Staff Officer, Miss Denis de Vitré, was appointed to assist H. M. Inspector of Constabulary and rose to the rank of Assistant Inspector of Constabulary. The Police Federation of England and Wales had long refused to admit policewomen to its membership and it was only after long arguments that women were allowed to join in 1948. The Oaksey Committee, set up in May of that year, examined the recruitment situation of the police force and confirmed that women had now been accepted as part of the force. However, it was not until 1974 that women police units were integrated with the male police forces.

(Lock, 1979; Tancred, 1951)

Women Sorters' Association

This Association, formed in 1906, was for Post Office sorters who had entered between the age of 15 and 18 by means of a competitive examination. It presented a petition shortly after it was established concerning wages and annual leave and in 1908 requested that the name of the grade be changed to Assistant (Women) Clerks. The same year, the Association approached the Fawcett Association (a Post Office organisation named after Henry Fawcett and having no connection with suffrage), with a view to amalgamation. After some initial opposition from the Postmaster-General, the amalgamation took place.

Women Teachers' Franchise Union

Formed in 1912, it campaigned for women's suffrage within the National Union of Teachers, attempting to overcome the considerable opposition of many male teachers. Its members were later active in forming the **National Union of Women Teachers**.

(A. J. R., 1913)

Women Writers' Suffrage League

Founded in 1908 to bring together women writers in the cause of women's suffrage, it included such prominent members as Sarah Grand, Cicely Hamilton, Beatrice Harraden, Marie Belloc Lowndes, Elizabeth Robins and Flora Annie Steel. Its secretary was Bessie Hatton. Its main activity consisted in propagandising suffrage, either in humorous pamphlets, often in conjunction with the **Artists' Suffrage League**, or writing agitprop plays in collaboration with the **Actresses' Franchise League**. It effectively ceased activity on the outbreak of war in 1914, although it was not officially wound up until 1918.

(Holledge, 1981; Whitelaw, 1990)

Women's Advertising Club of London

After the demise at the end of the First World War of the **Association of Advertising Women** which represented the interests of those employed in that profession, there was a hiatus. Five years later, in 1923, the Women's Advertising Club of London was established. Membership was by invitation only and consisted of 50 top executives in the United Kingdom with Marion Jean Lyon, advertising manager for *Punch*, as its first president.

Women's Advisory Council on Solid Fuel

This appears to have been formed in 1943 on government instigation, under the leadership of Mary Leigh, to complement the **Electrical Association for Women** and the **Women's Gas Federation and Young Homemakers**. Its object was 'to bring to the notice of the housewife the advantages of coal and other forms of solid fuel for domestic heat services and to advise her in the choice and operation of improved solid fuel burning appliances for space-heating, cooking and hot water'. It had close contacts with government ministries and with the Mining Association. In 1944 its chairman was the Hon. Lady Egerton, and its organising secretary was Mrs E. C. Rollo. From 1945 its address was 18 South Molton Street, London, W1. It was still active in 1959.

Women's Aerial League

The public interest in flying following the success of Wilbur and Orville Wright in the USA in 1903, John Moore-Brabazon, the first English aviator in 1908 and Blériot's flight from Calais to Dover in July 1909, led to a number of women enthusiasts flying, but at first only as passengers. The Women's Aerial League was formed to encourage and stimulate the invention of 'aerial craft and the things appertaining thereto'. To forward this venture, the League offered scholarships of £50 for three years to students of aviation at the College of Science and Technology, now Imperial College, London.

Women's Amateur Athletics Association

Games played little part in nineteenth-century girls' curriculum so the rise of athletics as a sport was slow. The Amateur Athletics Association was formed in 1880 by three Oxford University men with the intention of establishing a single body for English athletics. Although it was initially successful in this aim, the situation had changed by the early twentieth century. The popularity of athletics among women led to the decision to hold a *Réunion International d'Education Physique Féminine* at Monte Carlo in 1921 with teams representing European countries, followed later that year by an international meeting between England and France. With the success of the English teams, a Women's Association was formed in 1922 which became the Women's Amateur Athletic Association in 1926. It formulated the rules for women's athletics, organised and monitored women's athletics and offered advice on training. Universities and business houses rallied to the Association and there were soon some 23,000 members. The Association participated in the first Women's Olympic Games held in Paris in August 1922. At the first Women's International meeting held under its auspices in August 1924 at Stamford

Bridge before 25,000 spectators, Great Britain won against competition from Czechoslovakia, Italy, Belgium, France and Switzerland. By the 1950s Britain was the only nation retaining separate men and women's athletic associations.

(Webster, 1930)

Women's Amateur Rowing Association

The men's Amateur Rowing Association (ARA) was established in 1879 'with a view of forming representative British crews to compete against foreign and colonial clubs'. The majority of the controlling committee were university graduates. The Women's Amateur Rowing Association (WARA) was founded in 1926 and in the following year the Ladies' Boat Race between Oxford and Cambridge was held on the Isis over half a mile. The Association's objectives were to maintain among women the standard of oarsmanship as recognised by amateur rowing clubs; to hold a Regatta or Regattas if and when so determined by the committee, such regattas to be mainly or exclusively for women; and to promote the interests of boat racing generally among women. The Association suspended its activities during the Second World War, but returned with vigour thereafter. In 1948, its president was the Lady Desborough, its chairman Miss A. C. Gentry, its treasurer Mrs C. Wyndham-Fewing, its secretary Miss A. B. Freestone and the Regatta secretary was Miss J. Skinner. The first European championships were held in 1954. Both the ARA and the WARA had affiliated to the International Rowing Federation in 1947 and in 1961, the Women's Amateur Rowing Association was incorporated into the Amateur Rowing Association.

(Mason, 1989)

Women's Anti-Opium Urgency Committee

Apparently founded in 1908, its purpose was to help in abolishing the manufacture, sale and export of opium in and from India, except for medicinal purposes. Its honorary secretary was Miss R. B. Braithwaite, 312 Camden Road, London, N. It does not seem to have survived the outbreak of war in 1914.

Women's Army Auxiliary Corps

By the end of 1916, the War Office, because of the heavy casualties on the Western Front, decided to accept women for duties in the armies in Britain and France in a newly formed Women's Army Auxiliary Corps. Mona Watson, a medical doctor, was appointed as Commandant and Helen Gwynne-Vaughan as Chief Controller overseas. The War Office ruled that the Corps was a civilian organisation and that there would be no parity of ranks with the Army. Their headquarters at Devonshire House, Piccadilly, attracted many volunteers. By the end of 1916, some 57,000 women had been recruited, 10,000 of whom would serve in France. The first draft, consisting of cooks and waitresses, who reached the Front in March 1917, were joined by others who acted as typists, clerks, printers, bakers and signallers. In recognition of their work, the WAAC was renamed the Queen Mary's Army Auxiliary Corps in May 1918. The Corps ceased to exist after September 1921.

(Cowper, 1949; Gwynne-Vaughan, 1942)

Women's Automobile and Sports Association

It appears to have been active in 1931, when it offered accommodation to the **Women's Billiards Association** at 17 Buckingham Palace Gardens, London, SW1.

Women's Auxiliary Air Force

By August 1917, the War Cabinet had decided to form an independent air service, the Royal Air Force, from the existing Royal Flying Corps and Royal Naval Air Service, but it was not until 28 December that women began to be recruited. On 24 January 1918, the Air Council created the Women's Royal Air Force and members of the **Women's Army Auxiliary Corps** and the **Women's Royal Naval Service** working on stations were allowed to join the WRAF from 1 April 1918. The conditions of service differed from those of men; as enrolled rather than enlisted personnel, they were technically civilians in uniform, though members were more integrated into the main structure than the other two women's organisations. Amongst the many occupations were cooks, domestics, orderlies, clerks and storewomen. By August 1918, there were 15,000 members. The WRAF was finally disbanded in April 1920.

After the **Auxiliary Territorial Service** was set up in 1938, Royal Air Force Balloon Centres were at first often largely manned by members of the ATS. Differences in training needs were recognised and on 1 July 1939, the Women's Auxiliary Air Force was established with Jane Trefusis-Forbes as Senior Controller. By October 1943, the WAAF formed 16 per cent of the total RAF numbers and 22 per cent of the RAF personnel in Britain. In all there were approximately 250,000 WAAFs. As with the other services, the main function was to release men for operational duties. WAAFs undertook more than 50 occupations, including photographic interpretation, operation room work, radar, aircraft fitting, telegraphy, maintaining safety equipment and interpreting. Many served overseas, and of the 15 who volunteered for the Special Operations Executive, six died. Two members of the WAAF were awarded the George Cross. On 1 February 1949, the permanent service reverted to the First World War title of Women's Royal Air Force (WRAF). For the first time, rank parity was achieved with men. From 1 April 1994, the Service was integrated into the Royal Air Force.

(Escott, 1989; Taylor, 1988)

Women's Auxiliary to the Baptist Union of Scotland

Founded in 1909 to co-ordinate and organise work amongst Baptist women in Scotland and spread the gospel, this organisation holds regular meetings in local Baptist churches for fellowship, Bible study and prayers as well as for practical community projects. It promotes mother–toddler groups and visits and help for the disabled and the elderly. The local branches are organised into 11 district unions which meet and participate in rallies and national events. There are about 2,000 members in all. There is a quarterly magazine, *Newslink*.

Women's Billiards Association

It was founded in 1931 with the aims of encouraging, promoting and controlling women's billiards, and sponsoring championships and other competitions, team matches and tournaments. Its president was the Viscountess Elibank, the chairman, Mrs Longman and vice-chairman, Teresa Billington-Greig. Most of the Council members were also members of the **Lyceum Club**. Its headquarters were at 17 Buckingham Palace Gardens, London SW1, by courtesy of the **Women's Automobile and Sports Association**. It is uncertain when it ceased activity; in the 1980s a successor organisation coined the slogan 'A woman's right to cues.'

Women's Club

This early Club, situated at 9 Berners Street, Oxford Street, London, W, changed its name from Working Women's Club to Women's Club in 1872. It contained a reading room supplied with newspapers and periodicals, a conversation room and a visitors' room in which members could receive their friends, and a dining room which supplied refreshments at a moderate price.

Women's Committee Against War and Fascism *see* World Women's Committee Against War and Fascism (British Section)

Women's Co-operative Guild

The moving spirit of the Women's Co-operative Guild was Mrs Alice Acland, wife of the Liberal politician, Arthur Herbert Dyke Acland. The

Guild came into existence at the Congress of the Co-operative Union in 1883 with the title the Women's League for the Spread of Co-operation; it became the Women's Co-operative Guild in the following year. Until then, the Co-operative movement was predominantly run by men and the Guild was initially a group of mainly middle-class women who wanted to see this situation changed. Its general secretary was Margaret Llewelyn Davies. Starting with a membership of 51, by 1889 it had grown to 1,800 and by 1895 it had over 8,000 members with 196 branches, each with their own councils, and soon a majority of the membership was working class. Much of the early work was concerned with the establishment of clothing and coal clubs and visiting the sick, but the Guild's main aim was to introduce practical citizenship for women. It soon managed to place some of its members on representative committees of the Co-operative movement and urged members to play a full part in the management of their local Co-operative societies. It affiliated to the Labour Party in 1918.

In many respects, the Guild was more radical in its outlook than the Co-operative movement. During its history, the Guild, which attracted many more working-class women, conducted campaigns to establish minimum wages for shop workers, addressed the concerns of its married women members on such matters as maternity benefits, demanded votes for women and advocated divorce law reform. Between the two World Wars, especially during the secretaryship of Eleanor Barton (1921–37), the Guild was active in promoting peace, disarmament and international co-operation, high and stable levels of wages, and was involved in campaigns for increasing information on birth control and for abortion law reform. In addition, it encouraged women to stand for election to public bodies. After Eleanor Barton's somewhat authoritarian period as general secretary, the committee of the Guild took measures to limit the general secretary's powers. At the 1945 general election, several women were elected as Co-operative Members of Parliament. In 1963, the Guild changed its title to the Co-operative Women's Guild. Three years later, the Guild reached its peak with a membership of 62,000. It now devotes itself largely to charitable work and raising money for a range of community projects.

(Bailey, 1955; Gaffin and Thoms, 1983; Webb, 1927)

Women's Corona Society

The Corona Club was founded in 1900 to provide an annual occasion, usually a dinner, at which serving Colonial officers could meet, in a relaxed atmosphere, with the Secretary of State for the Colonies and officials of the Colonial Office. In 1937, a women's equivalent, the Women's Corona Club, was formed, with Lady Graeme Thomson as president, principally for the wives and widows of Colonial officers past and present. After the Second World War it increasingly became concerned with the welfare of families, and in 1950 it changed its aims to include this, and became the Women's Corona Society. Over the years its aims and membership broadened to include general Commonwealth issues. In 1990, its governmental core funding was finally withdrawn, but at the time of writing it continues to be active, and is based at the Commonwealth Institute, London.

Women's Council

A non-political and non-sectarian organisation set up in 1935, the Council exists to promote understanding between the women of Great Britain and Asia, and acts as a liaising agency between major Britain and Asian women's organisations. Activities include educational and cultural programmes and the provision of books to Bookaid International which operates centres in India and other Asian countries.

(Shreir, 1988)

Women's Cricket Association

Founded in 1926 following a match at Colwall, Worcestershire, between teams of hockey and lacrosse players seeking an alternative to tennis for a summer game. The game had been played at country houses in the last two decades of the nineteenth century and in 1892, the Original English Lady Cricketers had two professional teams. The chairman was Mrs Patrick Heron-Maxwell, a former president of the **All-England Women's Hockey Association**, a former organiser of the **Women's Land Army** and member of the National Executive of the **National Federation of Women's Institutes**. Its aim was to form clubs so that an number of county teams could be established. The Association arranged tours and fixtures in different parts of the country for colleges and public schools who were already offering the sport. By the end of the 1922 season, the Association consisted of 347 individuals, ten clubs, 28 schools, six colleges and two business houses. The first public match took place in July 1929, a one-day fixture between London and District and the Rest of England. County Associations were formed in 1922 consisting of Middlesex, Lancashire, Kent, Nottinghamshire and Somerset, joined later by Yorkshire. In 1934, the first test series against Australia was played at Brisbane and England won by nine wickets. By 1936, 120 clubs and 100 schools had affiliated to the Women's Cricket Association.

(Joy, 1950; Williams, 1989)

Women's Defence Relief Corps

It was founded apparently in 1914 by Mrs Dawson Scott, with

headquarters at 6 King Street, Southall, Middlesex (there was also a 'Manchester initiative'). Its patron was Miss Eva Fenton, 2 Albermarle Street, London, W1. Its aims were to undertake civilian or semi-military work which would enable more men to enlist, and to develop women's character in order to make them into better citizens. The civilian part of the work was intended to consist of literally taking over men's jobs, at men's wages, but giving an undertaking ('preferably on paper') that the jobs would be given up to the men on their return from the war. The semi-military side involved training in drill, marching, signalling, scouting and the use of arms with the aim of releasing men from such duties as the guarding of railways and public buildings and doing picket work in co-operation with civic authorities and the police. Corps members were called Helpers, and wore a scarlet H as a badge. The age range was 16 to 50. In fact, this organisation appears to have been short-lived.

Women's Education Union

Begun in 1871 as the National Union for the Improvement of the Education of Women of All Classes, it soon became known as the Women's Education Union. The moving force was Maria Grey who was supported by, amongst others, Emily Shirreff, Mary Gurney and Lady Stanley of Alderley. She had addressed the **North of England Council for Promoting the Higher Education of Women** where she had met a number of men and women concerned about female education, and later that year, in November 1871, launched the WEU. Maria Grey was elected its first president but relinquished the post to become its organising secretary. The Union had several aims. It wished to improve the qualifications of women teachers in order to achieve higher standards with their pupils. Teacher-training facilities were required with examinations offered by a recognised authority. It hoped that existing grammar schools would form classes for girls taught by regular teaching staff and endorsed the need for adult and university education for women. The Union also provided loans and scholarships for women students. Most important of all it set out to found new, academically, sound, low-fee days schools which had to be launched as commercial ventures. The Girls' Public Day School Company Ltd, later the **Girls' Public Day School Trust**, subsequently came into being. The Union started a Teachers' Training and Registration Society in 1876 and three years later opened its first institution, the Maria Grey Teachers' Training College.

(Bolt, 1993; Ellsworth, 1979)

Women's Educational and Industrial Union of London

This was active in 1895, and its activities were reported in the periodical *Shafts*.

Women's Emancipation Union

This described itself as an 'association of workers to secure the political, social and economic independence of women'. Although its activity seems to have begun in 1891, it was only formally established on 24 October 1892 at an inaugural meeting at the Midland Institute, Birmingham, called by Elizabeth Wolstenholme Elmy after she had been expelled from the **Women's Franchise League**; other prominent members of the League who joined her included Mona Caird, Lady Florence Dixie, Harriet McIlquham, Charlotte Carmichael Stopes and Frances Swiney. It was active in producing pamphlets urging the cause of women's enfranchisement. Its final meeting took place on 1 July 1899.

(Joannou and Purvis, 1998)

Women's Emergency Corps

It was started within two days of the declaration of war in 1914, with the Duchess of Marlborough as treasurer and the Hon. Mrs Evelina Haverfield and Lena Ashwell as secretaries (Beatrice Harraden was also involved), and was based at Old Bedford College, 8–10 York Place, Baker Street, London, W. Its purpose was to gather together and classify all expert voluntary workers and to help societies 'of all grades of opinion' co-ordinate efforts to cope with the war situation. It also aimed to form an industrial centre for organising paid employment particularly for 'middle-class workers thrown out of work by the war'. It lasted at least long enough to produce an annual report in 1915. It is unclear to what extent, if at all, it interacted with the Women's Service Bureau of the **London Society for Women's Suffrage**.

Women's Emigration Society

This was founded in 1880 mainly on the initiative of Louisa M. Hubbard and Mrs Caroline Blanchard, who were aware of the problems encountered by the **Female Middle-Class Emigration Society**, and had ideas on how to deal with them. Contemporary accounts of this Society are to be found in Louisa Hubbard's periodical *Work and Leisure*. Unlike the Female Middle-Class Emigration Society, it stressed that the sort of work emigrant women were most likely to find was of a domestic nature. It ceased functioning in 1884, most of its active members having already formed the **Colonial Emigration Society**; the rest joined the **United British Women's Emigration Association**.

(Hammerton, 1979; Monk, 1963)

Women's Employment Defence League

The League was founded in 1895 to protest against the provisions of the

Factory and Workshops Bill of that year, relating to the employment particularly of children, which was characterised as restrictive legislation. Its president was Jessie Boucherett, the chairman was Ada Heather-Bigg, and the secretary was Eleanor White. It was based at 4 Frith Street, Soho Square, London. It does not appear to have been long-lived but was the foreunner of the **Freedom of Labour Defence**.

Women's Employment Federation

Formed in 1933 as the National Federation of Organisations concerned with the Employment and Training of Women, it carried on and expanded the work formerly done by the Women's Service Bureau of the **London and National Society for Women's Service**, with which it maintained close connections and initially shared premises. It brought together representatives of university-appointments boards, schools, colleges and professional organisations to provide a careers advisory service especially tailored to the needs of women. It also had a loan fund which lent money free of interest to cover training likely to lead to useful and progressive work. It was funded by subscriptions from schools, universities and other interested organisations. It also produced annual careers guides for women, as well as occasional publications directed at special groups of women (e.g. older women, arts graduates, etc.). Its first organising secretary was Ray Strachey (Mrs Oliver Strachey); in 1939 it secured the patronage of the then Queen (Queen Mother at the time of writing). On the outbreak of war the Federation moved its offices to Bedford College in Regent's Park, London, and thereafter increasingly separated its activity as well as its address from that of the London and National Society for Women's Service. In 1971 it changed its name to the National Advisory Centre on Careers for Women, and in 1991 to Careers for Women. It ceased activity in 1995.

Women's Employment Publishing Company *see* Central Employment Bureau for Women

Women's Engineering Society

This Society was founded by Lady Parsons early in 1919 at 46 Dover Street, London, W1, to build on and extend the opportunities that had opened to women in the course of the First World War. The original officers were Caroline Haslett as general secretary, and Lady Parsons as president. It soon had branches outside London, beginning with Newcastle, Kirkcudbright, Manchester and Aberdeen. Between the wars it had a number of prominent presidents, most notably the aviator Amy Johnson (1934–37). Shortly after its establishment, it moved to 26 George

Street, Hanover Square, London, W1, and changed addresses frequently; at the time of writing it is based at the Department of Civil Engineeering at Imperial College, London. Its objectives continue to be to promote awareness of engineering as the prime creator of wealth in our society, and the contribution women can make to it; to promote the education and training of women engineers; to ensure that women engineers can influence the process of policy formation and decision-making in government and other organisations; to foster good practice in the employment of women engineers in order to enable them to progress equally with their male colleagues; and to provide a network of members and support. Its organ since 1919 has been the *Woman Engineer*.

(Messenger, 1967)

Women's Farm and Garden Association

This Association was established in 1899 to promote the study of agriculture and horticulture among professional women engaged in this work in the United Kingdom and overseas. It assisted members in obtaining employment and in disposing of produce, mainly through shows. Advice was available to school leavers and students intending to take up agriculture as a career. The Association was in touch with the **Society for the Overseas Settlement of British Women** and had the latest information with regard to openings for women landworkers in all the Dominions. Members met for discussion of common problems and links established with other organisations, including membership of the Royal Horticultural Society, the United Nations Women's Advisory Committee and the **National Council of Women**. The organisation had changed its title from the Women's Farm and Garden Union with its headquarters first at 45 and 46 Queen Anne's Chambers, Westminster and later at Courtauld House, Torrington Square, London, WC1, which had been donated by an enthusiastic member of the Association, Miss Katherine Courtauld. Its president was HRH the Princess Louise, Duchess of Argyll. In 1916, the WFGA set up a voluntary Women's Land Corps which the government took over as the **Women's Land Army** in the following year.

(King, 1999)

Women's Forum *see* Women's Group on Public Welfare

Women's Franchise Declaration Committee

This was formed by Clementina Black in November 1906. Its main stated aim was to collect signatures to a declaration by professional and business women in favour of women's enfranchisement. It was also responsible for producing the periodicals *Women's Tribune* and *Women and Progress* in 1906 and 1907.

(Crawford, 1999)

Women's Franchise League

Formed in 1889 largely to protest against the negative attitude towards married women of the major women's suffrage societies, its initial membership included Emmeline and Richard Pankhurst, Elizabeth Wolstenholme Elmy, Josephine Butler, Jane Cobden, Ursula Bright, Harriot Stanton Blatch, Harriet McIlquham, Alice Scatcherd and Mrs P. A. (Clementia) Taylor. Personal tensions within the League led to a split in 1891, resulting in the formation of the **Women's Emancipation Union**. The League seems to have made up for its small membership (probably only a few hundred) by its vocal activism. However, its major supporters fell away in the late 1890s and it ceased functioning some time after 1897.

(Holton, 1996; Joannou and Purvis, 1998; Pankhurst, 1931)

Women's Freedom League

The League began as a breakaway from the **Women's Social and Political Union**, with a more democratic structure, originally led by Charlotte Despard, Teresa Billington-Greig and Edith How-Martyn. At first it maintained militant action, but together with the Women's Social and Political Union it declared a truce in 1910; unlike the Women's Social and Political Union, it did not resume militancy thereafter, but campaigned for the vote by constitutional means, as well as for a range of feminist causes, such as equal pay, equal entry to the professions, equality in public life, ending sexual abuse of children, etc. During the First World War it continued campaigning for the vote. From 1919 to 1928 it worked actively for equal franchise, among many other issues, and continued thereafter as an active feminist organisation until 1961. Its members were very diverse: Catholic women seem to have been disproportionately prominent, but equally some of its leading members (for example, Edith How-Martyn) were vigorous campaigners for birth control. This diversity led to a number of surprising silences on matters where it was felt that the membership might be split. In 1909 the League started its periodical the *Vote*; this was effectively subsidised by Dr Elizabeth Knight, and ceased abruptly when she died intestate in November 1933. In 1934 it was succeeded by the much more modest *Women's Freedom League Bulletin*, which continued in cyclostyled format until the demise of the League.

(Linklater, 1980)

Women's Gas Council

It was formed in 1935 as an independent national organisation to meet the need for a body through which the large numbers of women who used gas in their homes might voice their opinions and co-operate with similar organisations working for improvements in home life. The Council's aims were to help improve home life by the distribution of information on scientific home management; to assist women to study homecraft and the latest developments in design, kitchen-planning and smoke abatement with particular reference to gas; to establish branches where women could meet socially for free discussion and exchange of views on all subjects referring to home-making; and to provide an avenue of approach along which the views of housewives could be conveyed to the appropriate authority. In 1951 it changed its name to the Women's Gas Federation.

The Council worked in close co-operation with government departments and local authorities. By the late 1950s it had 135 branches. Its first president was Lady Londonderry, the chairman was Mrs Wolseley-Lewis and the organising secretary was Miss Kathleen M. Halpin, who later became president. The national headquarters were originally at 20 Grosvenor Place, London, SW1, and subsequently at 5 Grosvenor Crescent, London, SW1. Its organs were *Fanfare* (1936–47) and *Commentary* (1948–67). It also produced a series of quarterly illustrated leaflets entitled *I pass this on to you*. It affiliated to the **National Council of Women**, the **National Women Citizens' Association**, the **Women's Employment Federation** and the League of Nations Women's Advisory Council. In addition, Miss Halpin was elected as representative to the **Soroptimist** Club of Greater London. The Federation was still active as the Women's Gas Federation and Young Homemakers in 1977.

Women's Gas Federation and Young Homemakers *see* Women's Gas Council

Women's Group on Problems Arising from Evacuation *see* Women's Group on Public Welfare

Women's Group on Public Welfare

In September 1939, the National Council of Social Service called a conference for representatives of organisations concerned about the consequences of wartime evacuation of women and children. As a result, the Women's Group on Problems Arising from Evacuation was formed under the chairmanship of Margaret Bondfield, with the National Council of Social Service providing secretariat and accommodation. In 1940, the

Group's name was changed to the Women's Group on Public Welfare, to reflect its widened scope. Its main aim now was 'to bring the experience of its constituent organisations to bear on questions of public welfare, more especially those affecting women and children'. Representatives were drawn from national women's organisations and other voluntary organisations with a large women's membership. The Group had no individual members, only representatives of its contributory organisations, and action could only be taken in the name of those organisations agreeing to it. The main work of the Group was mainly done by sub-committees and working parties, which undertook a number of surveys, and established links with foreign organisations. In 1942, the Group co-operated with the local Group Action Councils of the Federation of **Soroptimist Clubs** to form the **Standing Conference of Women's Organisations**. Its activities continued after the War, until in 1975 the National Council of Social Service restructured, and changed its name to the National Council of Voluntary Organisations; the Group also changed its name to Women's Forum. When the National Council of Voluntary Organisations could no longer continue its secretarial and financial support of the Women's Forum, the latter was wound up at its annual general meeting in 1980.

Women's Guild of Empire

It was founded by Flora Drummond and Elsie Bowerman in 1924 or 1925, with the aim of promoting co-operation between capital and labour, preventing strikes and combating Communism (though it also disapproved of Fascism). Flora Drummond was the Controller-in-Chief, and the Guild appears to have lost most of its impetus after her death in 1949. Some time subsequently it appears to have merged with the League of Empire Loyalists, one of the bodies which came together in the 1960s to form the National Front.

Women's Home Defence

An organisation set up by Dr Edith Summerskill, MP, in 1940, to teach women how to use firearms in the event of invasion. Open to women of all ages, it was initially treated with scepticism by the press, but it gradually gained acceptance By 1942, there were 250 units in England consisting of 20,000 women. It was supported by high-ranking officers of the three Services, including Admiral of the Fleet Lord Cork of Orrery and Marshal of the Royal Air Force Sir John Salmond. There was a campaign to make this an official part of the Home Guard, but this was resisted vehemently by the government.

(Goldsmith, 1943)

Women's Imperial Defence Council

The Council, described as a 'non-political body', is only known from a report of a meeting at the Cannon Street Hotel, London in 1917, chaired by General Sir Arthur Turner, and addressed exclusively by men, including the future Home Secretary, Sir William Joynson-Hicks. The object of the meeting was to expose the extent of German influence in Britain.

Women's Industrial Council

Established in 1894 'to watch over the interests of women engaged in trades, and over all industrial matters which concern women'. A wholly unsectarian organisation independent of political parties, its leading members included Clementina Black, Margaret Bondfield, Margaret MacDonald, B. L. Hutchins and Catherine Webb. The secretary, Mary Macarthur, filled the post for seven years. The Council aimed at remedial legislation which would limit the hours of women's work, promote better health and safety regulations and the recruitment of more Factory Inspectors. One of the Council's tasks was to report breaches of Factory and Workshop Acts to Inspectors. Its main approach was to investigate women's work conditions, scrutinise public bills and reports which offered better working conditions and organise petitions. In 1897, the Councils' Investigation Committee enquired into the effects of factory legislation on women's employment and, in 1900, successfully co-operated with other organisations in making the government withdraw its Factory Bill. The Council's Legal Committee had the previous year drafted a Bill for the better regulation of home work. It was introduced in Parliament several times between 1899 and 1907. Members were divided on the issue of Trade Boards and licensing as means of remedying the problem of sweated labour. The right of girls to technical and trade training was raised and the Council was also concerned with female unemployment. In 1905, it suggested that domestic service could be a suitable outlet, though it recommended that training should be given to prospective candidates. The house journal was the *Women's Industrial News*.

(Mappen, 1986)

Women's Institute

Not to be confused with the later **National Federation of Women's Institutes**, it was founded in 1897 at 15 Grosvenor Crescent, London,W, by Mrs Nora Wynford Philipps. It originally seems to have acted as an academic and professional branch of the **Grosvenor Crescent Club**, whose premises it initially shared, and acted as a centre where women engaged in the fields of educational, social and philanthropic work could meet. The Institute assisted members requiring employment and carried out professional and academic research on their behalf. It contained a library (with 2,500 volumes in 1898) and an information bureau, and in 1898 it published a *Dictionary of Employments Open to Women.* It organised lectures and also ran courses, notably a secretarial training course. At this time it had over 800 members; the annual subscription was one guinea, or 10s 6d for professional members. In 1900, it moved to 92 Victoria Street, London, SW1. In 1925 it again moved, to 70 Victoria Street, and appears to have ceased activity shortly thereafter.

(Crawford, 1999)

Women's Institutes *see* National Federation of Women's Institutes

Women's International Art Club

This Club was founded in 1898 to enrol women painters of different nationalities and to help them exhibit in London and elsewhere. It was formed in Paris by a small coterie of women painters led by Miss M. A. Hurst, Miss Spence Bate, Miss Alice Hay, Miss Haig and Miss Badnell. After its first exhibition at the Grafton Galleries in 1900 in London, the Club adopted its title of the Women's International Club. At first, members were required to have studied in Paris, but this rule was soon abolished. It stated that 'with regard to the treatment of Art, the one aim and object has been and will be, that the Club be representative of no particular school of painting, but exhibit fairly all schools, as all schools are represented in Paris. Good drawing to be the chief test required of members.' The normal requirement for membership was that the candidate should have exhibited twice in the three years previous to joining. Art was not confined to painting: members exhibited sculpture, etchings, colour prints and wood engraving as well as embroidery, enamel-, jewel- and leather-work. Exhibitions of members' work became an annual event. The Club had no official meeting place and was dependent on members to provide accommodation. In 1908, lay members who were interested in the arts were admitted for the first time. The first secretaries were Miss Spence Bate and Miss Ida Sheridan; they were succeeded in 1899 by Miss Evelyn Howard.

Women's International Maybrick Association

This Association was founded at a meeting at the **Pioneer Club** on 30 May 1895 to campaign for the release of Mrs Florence Maybrick, an American woman who, in 1889, was imprisoned for allegedly murdering her husband James (possibly 'Jack the Ripper'). She was sentenced to death but subsequently reprieved. The Association aimed to publicise 'the grounds for believing that Mrs Maybrick had an unfair trial, and is suffering unjust imprisonment'. A committee was formed to assist in opening a popular subscription of one shilling each. The president was Helen Densmore, and the secretary was Mrs M. M. Cook of 10 West Chapel Street, Mayfair, London. The Association may have established links with a similar organisation in the United States, the committee of which included Elizabeth Cady Stanton. It was not until 1904, some 15 years after the trial, that Mrs Maybrick was released from prison.

Women's Labour League *see* **National Labour Women's League**

Women's Land Army

The first Women's Land Army was set up in 1917 when, because of the German Navy, there was only about three weeks' supply of food left in Britain. By the end of the First World War, there were 23,000 women who took on the jobs of milkers, tractor drivers, field-workers, carters, ploughmen, thatchers and shepherds. The WLA was disbanded at this time. However, War Agricultural Committees, formed in 1936 when the international situation looked serious, calculated that by 1940 there would be a shortfall of some 50,000 agricultural workers. To fill this gap, the Women's Land Army was relaunched on 1 June 1939. Lady Gertrude Denman, who had been involved in organising the original WLA, was appointed Director. Each county was responsible for recruiting and interviewing volunteers, though they could subsequently be sent to other areas where help was needed. One-third of the recruits were from London or industrial towns in Lancashire and Yorkshire. Their tasks were similar to those carried out by their predecessors, and included milking, hedging, haymaking, threshing, land reclamation and forestry (see **Women's Timber Corps**). By 1943, apart from the Timber Corps, there were 43,000 Land Girls. The writer Vita Sackville-West took a close interest in the work. She was District Representative for a large area of Kent and wrote the official history, *The Land Army* (1944). A monthly publication was issued, called the *Land Girl*, and edited by Margaret Pyke. A Scottish Women's Land Army was also set up. The WLA was finally disbanded in 1950.

(Tyrer, 1996)

Women's League for Municipal Reform

This organisation, with its headquarters at 2 Bridge Street, Westminster, London, and closely associated with the London Municipal Society, aimed at opposing Socialism at municipal and national levels and encouraging women to maintain an interest in the conduct of business by different local authorities.

Women's League of Health and Beauty

This self-supporting commercial enterprise was established by Mary Bagot Stack in 1930, and continued by her daughter Prunella Stack. Originally called the Build the Body Beautiful League, it was designed 'for business girls and busy women to enable them to conserve and improve their physique'. Its original stated aim was 'Racial Health' (referring to the betterment of the human race), but this was later modified to avoid any association with theories of white or Aryan supremacy; in practice it has been multi-racial. It was particularly noted for its 'uniform' of white blouse and black knickers. It inspired a host of commercial competitors, as well as being a precursor of the late twentieth-century gyms and health clubs; unlike these, however, and more in keeping with its contemporary organisations in nationalist and Marxist regimes (to say nothing of Busby Berkeley films), it also arranged mass gymnastic displays. Although still in existence as late as the 1990s, its real heyday was in the 1930s. Its organs were *Beauty* and *Mother and Daughter.*

(Matthews, 1990)

Women's League of Unity *see* **Women's National League of Unity**

Women's Legion

During the First World War, when the Army was experiencing a severe shortage of manpower in 1915, Lady Edith Londonderry became the founder and Director-General of the Women's Legion. Members wore a khaki uniform with a badge depicting the bronze figure of Victory holding a laurel wreath (called by many of the Legion 'the lady with the frying pan') and the motto *Ora et Labora*. The Legion recruited women into one of four sections, Agricultural, Herb, Canteen and Ambulance and Mechanical Transport. The latter supplied drivers for the Royal Army Service Corps and the first cookhouse was opened at a convalescent camp at Dartford in August 1915. The Legion was officially recognised by the Army Council in February 1916. Members were recruited through the labour exchanges and were paid by the War Office. Personnel differed widely in their qualifications, ranging from amateurs to those with domestic diplomas and degrees. The Women's Legion Canteen section, consisting of some 30,000 members, was integrated into the **Women's Auxiliary Army Corps** in 1917. During the Second World War, the

Legion was closely involved in helping the civilian population on the home front.

(Colclough, 1920; de Courcy, 1992)

Women's Liberal Federation

By the 1880s, many Liberal women demanded a voice in political and social affairs. The establishment of the **Primrose League** by the Conservative Party in 1884 was countered by the formation into a Federation of the already existing Women's Liberal Associations throughout the country. The first Women's Liberal Association had been set up in Birmingham and was followed by the first official Association at Bristol in 1880 and by others in Darlington and York. By 1886, there were 15 WLAs throughout the country. A central body was considered necessary and at a meeting at Queen Anne's Gate, London in May 1886, a resolution was passed to form all existing WLAs into a Union. The inaugural meeting of the Women's Liberal Federation was held in February 1887, by which time there were 40 WLAs with more than 10,000 members. Catherine Gladstone was its first president. The objects of the Federation were to promote the adoption of Liberal principles in the government of the country, to promote just legislation for women and to protect the interests of children, advance political education by meetings, lectures and the distribution of literature and to establish a WLA in every constituency.

The policy of the WLF was formulated at its annual Council meeting, held in London in May or June. One of the most important issues was that of women's suffrage. William Gladstone and other leading Liberal politicians opposed the move, and this led to a split in the Federation. In 1892, Catherine Gladstone resigned her post as president and the WLF became known as the Women's National Liberal Federation, though the 'National' was dropped from its title in 1935. The breakaway body, the **Women's National Liberal Association**, was formed by 11 ladies from the Executive of the WLF who believed that women's franchise, not being in the Liberal programme, was inconsistent with their position as Liberal women. On the other hand, some members felt that the WLF was not sufficiently active in promoting the suffrage cause, and formed the **Union of Practical Suffragists**.

The WLF flourished under the presidency of two dedicated presidents, the Countess of Aberdeen and the Countess of Carlisle. The latter, a celebrated temperance campaigner known as the 'Radical Countess', was a fervent Liberal and an activist for social reform. Her 14 years as president saw the Federation flourish. From its beginnings, with 16,000 members, by 1908 there were 100,000 and approximately 1,500 WLAs.

The Liberal victory in the general election of 1906 owed much to the efforts of the WLF. Up to the First World War, suffrage was one of the main issues, but Asquith's stance on the subject disappointed many members, and some broke away. From the 1920s, the WLF supported Eleanor Rathbone, then an Independent MP, in her campaign for family allowances and pressed for divorce law reform, separate taxation for married women and better widows' pensions. In the 1930s, the Federation endorsed plans for peace, disarmament and internationalism, mainly through the endeavours of Margery Corbett Ashby. In the 1950s, the WLF was involved in the pursuit of equal opportunities and equal pay and continued to give help during general elections. In 1988, the WLF was wound up and was succeeded by the Women Liberal and Social Democrats; the title was later shortened to Women Liberal Democrats.

(Pugh, 1992; Tooley, 1897)

Women's Liberal Unionist Association

The introduction of a Home Rule Bill in 1886 and its subsequent defeat in the Commons led to a split in the Liberal Party. The Association was set up in May 1888 with the aim of upholding the legislative union between Westminster and Ireland and it opposed the Home Rule Bill of 1893. The Association's main objective was to attract voters to the cause by means of canvassing, debates and public meetings. It attracted many talented speakers for the Unionist cause, including Isabella Tod, Lady Frances Balfour and Millicent Fawcett. An interesting feature was the series of discussion meetings held in London each autumn and winter. Lady Stanley of Alderley was its first president. By 1891, it had recruited over 5,000 members, of whom about a third were from the Birmingham area where Joseph Chamberlain was MP. Two years later it was resolved to affiliate mixed associations of Conservative and Liberal Unionists though without giving them voting power on the Association's Council. The number of members had risen to 15,700 women in 60 branches by 1895. Nevertheless, the Association remained small compared with the **Women's Liberal Federation**. Its impact on the political scene was modest but through its candidates several members forged links with other parts of the women's movement, especially the suffrage cause.

(Shiman, 1992)

Women's Local Government Society

The 1888 County Council Act seemed to promise that women would be eligible to be elected as members of the new County Councils. In November of that year, 12 London-based Liberal women met to form the Society for Promoting Women as County Councillors. In 1893, its name was changed to the Society for Promoting Women to all Local Government Bodies and soon after to the Women's Local Government Society. One of its main movers was Amie Leigh Browne, who also made substantial financial contributions to the Society. Millicent Garrett Fawcett was its treasurer for two years. When two Liberal candidates, Lady Margaret Sandhurst and Jane Cobden, were elected to the London County Council, the government stated that it was never their intention that women should be eligible and this was confirmed in a subsequent law case (Beresford Hope *v.* Sandhurst, 1889). The Women's Local Government Society mounted a campaign aimed at redressing this anomaly and a Bill was presented in the Commons to reverse this ruling but without success.

During its existence, the Society sponsored independent women candidates, avoiding party labels and although feminist in its outlook, eschewed the suffrage cause. It continued to work to elect women councillors by promoting parliamentary legislation, by lobbying MPs and by distributing leaflets and petitions on a large scale. The 1894 Local Government Act allowed women to be elected to London vestries as Poor Law Guardians, and the Parishes and District Council Act of the same year allowed them to be councillors in parishes, urban and rural districts. The return of a Liberal government to power in 1905 led to a further campaign with Susan Lawrence a prominent figure. In 1907, an Act became law allowing women, regardless of marital status, to seek election to borough and county councils, providing they were ratepayers. Up to the First World War, the Society scrutinised legal rulings to ensure that no new impositions were placed on women's rights, and it promoted the appointment of women to posts in local government. By 1920 in England and Wales there were 320 women councillors serving on town and county councils and a further 300 on urban and rural district councils. It ceased to exist in 1925.

(Hollis, 1987)

Women's London Gardening Association

Mentioned in the *Queen* in 1891, the Associaiton appears to have been a sort of florist business. It was based at 62 Lower Sloane Street, London, SW.

Women's Migration and Overseas Appointment Society *see* **Society for the Oversea Settlement of British Women**

Women's Missionary Society *see* **Church Missionary Society: Ladies' Union**

Women's Municipal Party (1)

Formed in 1913 by the Duchess of Marlborough, who became its chairman, the Party aimed at attracting women, irrespective of political leanings, to stand for election as Councillors at municipal elections. Candidates were expected to adopt the election policy of the Party's Council regarding women and children, though in other matters they would be free, if elected, to vote as they wished. The Association, which hoped to mobilise the women's vote, attracted women with a range of political opinions. A number of members of the **Women's Local Government Society** joined the Women's Municipal Party; at one stage its Executive included such diverse personalities as Eleanor Rathbone, Mrs Pember Reeves and Lady Frances Balfour. Independent local government women were regarded as a nuisance by the main political parties and the organisation faded away after the First World War.

(Hollis, 1987)

Women's Municipal Party (2)

It was founded in 1913 by a Mrs O'Sullivan with the aim of supporting women candidates irrespective of party policy, on the grounds that 'the lady Councillors now sitting on the London County Council cannot be reckoned with a solid vote on special women's questions, as they have been run on party lines by their several political organisations'. It appears to have had no connection with the preceding entry. It does not seem to have lasted into 1914.

Women's National Anti-Suffrage League

Formed in 1908 under the leadership of Mrs Humphry Ward and Violet Markham, it opposed the admission of women to the parliamentary franchise, while supporting their representation in municipal and local government in general. Its paper was the *Anti-Suffrage Review*. In 1910, it was merged with the **Men's League for Opposing Woman Suffrage** in the **National League for Opposing Woman Suffrage**.

(Harrison, 1978)

Women's National Co-operative Self-Help Society

This was founded about 1895 with addresses at 61–62 Chancery Lane and 7–8 Southampton Buildings, London, WC; it seems to have been an initiative, probably the idea of a man, to stimulate women's producer co-operatives in order to commence a system of distributive trading 'without shops', as well as giving discounts by buying at wholesale prices. It does not appear to have elicited any response.

Women's National Health Association

In 1906 Lady Aberdeen, after the reappointment of her husband as Lord-Lieutenant of Ireland, started this Association in that country mainly to combat the spread of tuberculosis. The work initially was largely carried out by means of a touring entourage staffed by nurses armed with information on health and hygiene. Within four years, over 150 branches had been formed.

(Luddy, 1995)

Women's National League of Unity

Also referred to as the Women's League of Unity, this is only known in connection with the attempts in 1938–39 to start a women's newspaper, variously entitled the *Woman's Newspaper,* the *Woman's Daily Newspaper* and the *Woman's National Newspaper*. Its membership seems to have consisted of former militant suffragists; Emmeline Pethick-Lawrence and Flora Drummond were especially visible.

Women's National Liberal Association

In 1892, when the **Women's Liberal Federation** accepted women's suffrage as one of the objects of the Federation, a number of prominent members, including Lady Fry and Lady Hayter and nine other members of the Executive, resigned and formed a rival organisation, the Women's National Liberal Association. The first meeting was chaired by Sir Arthur Hayter, who suggested that the Association located its office near to the male-dominated National Liberal Federation and use their literature. Catherine Gladstone at the same time resigned as president of the Federation though she did not join the WNLA. Members were not necessarily against women's suffrage, but protested at the inclusion of an 'object' which was not part of the original constitution of the Federation. The new organisation looked to 'the preservation of absolute liberty of local action on the part of all members and Association in dealing with questions about which there is a difference of opinion within the Liberal Party'. It pledged to support the Party by educational work and political propaganda. The Association was not successful. There were too few experienced women speakers to promote the objects of the Association, which were lacking in definite purpose, and it failed to attract membership.

Women's Navy League

This appears to have been a women's affiliate of the Navy League. In the *Englishwoman's Year Book* for 1910, its secretary is given as Mrs W. Rowland Fisher, 7 Castelnau Road, London, W.

Women's Party *see* **Women's Social and Political Union**

Women's Peace and Arbitration Auxiliary

This Society, formed in April 1874, was an Auxiliary of the London Peace Society, established in 1816, and was affiliated to it. It contended that 'War is inconsistent with the spirit of Christianity and with the true interest of mankind.' The object of the Women's Peace and Arbitration Auxiliary was to adopt such practical measures as lay within the sphere of women to promote peace.

Women's Press Association

This Association was started about 1894 by a small syndicate of trained women journalists to supply literary contributions to newspapers, dealing with all subjects which might be of interest to women. The Association was intended to combat the system whereby the large agencies employed men only to collect and distribute information. The editor was Miss Drew and the Association was located at 35 Hastings House, Norfolk Street, Strand, London.

Women's Press Club

The Club was started at a meeting of women journalists called by Phyllis Deakin of *The Times* on 1 April 1943 at the Falstaff pub in Fleet Street, attended by 62 women, to form a Women's Press Club along the lines of the Press Club, to which women were not admitted. A need was felt for networking (most of the women at the meeting were unknown to each other, despite the fact that they all worked in Fleet Street), which included premises to provide a meeting place, especially for talks, and to offer hospitality to distinguished foreign visitors. The premises were acquired at 52 Carey Street, London, WC2, and the first annual general meeting took place there on 31 October 1944, chaired by the president, Lady Rhondda, with Phyllis Davies as vice-chairman and Hilda Grosvenor-Gill as secretary. It also published a newsletter, the first editor of which was Ann Temple. Other prominent members in succeeding years included Julia Cairns, Marjorie Proops and Mary Stott. It worked successfully in the 1940s and 1950s, but in the 1960s it began to experience severe financial difficulties, with the result that in 1968 it had to leave its Carey Street premises. It was wound up in 1972.

(Deakin, 1984)

Women's Printing Society

Formed in 1874 by Emma Paterson, possibly with the involvement of Emily Faithfull, it acted both as a printer for women's organisations and periodicals, and as a training organisation for women wanting to be apprenticed both to the printing trade and to such related occupations as proof-reading, occupations in which there was fierce resistance to women's involvement. Initially, £350 was raised in capital with shares of £1,100 subsequently taken up. By 1889, 17 young women were employed by the Society, learning all phases of the work. It provided technical training for young women and paid wages at the trade rate. Their share in profits was in proportion to their wages. Sums distributed ranged from five shillings to £13. It originally had premises at 21b Great College Street, London, SW, but in 1893 it moved to 66–68 Whitcomb Street, and in 1908 to 31–35 Brick Street, Piccadilly, London, W; at this time it employed 33 women, three men and six boys. It was still active in 1916, but does not appear to have survived the First World War.

(*Women's Printing Society*, 1913)

Women's Progressive Society

Apparently founded in 1890, and closely related to the **Union of Practical Suffragists**, its organ was originally the *Iris*, although later reports on it were published in *Shafts*, whose editor, Margaret Shurmer Sibthorpe, was a prominent member, as was Mrs Massingberd of the **Pioneer Club**. Its general aims were the emancipation of women, especially by the parliamentary vote. It seems still to have been in existence in 1897.

(Crawford, 1999)

Women's Protective and Provident League *see* Women's Trade Union League

Women's Protestant Union

It was founded in 1891 'to unite women of every class and denomination in maintaining and defending, by individual and combined prayer and effort, Protestant Christianity, and in resisting the encroachments of Romanism and Ritualism, which are undermining the spiritual, moral and social liberties of our country'. Its annual reports bore on the cover the Union flag, with various biblical mottoes, of which the first and most consistent was 'Earnestly contend for the faith' (Jude 3). The president and honorary secretary, presumably also the founder, was Mrs W. Reirson Arbuthnot (M. H. Arbuthnot), of Plaw-hatch, East Grinstead, Sussex until her death in 1912, when the presidency was taken over by the Lady Agnes A. Anderson. From 1903, its address moved from East Grinstead to 20 Exeter Hall, Strand, London, WC, and in 1907 to 26 Denison House, 296 Vauxhall Bridge Road, London, SW. The minimum annual subscription was five shillings. In the early 1900s its membership remained in excess of 1,350. Its organs were *Protestant Woman* and *Young Protestants.* It was still active in 1914; the youth section was still in existence in 1948.

Women's Publicity Planning Association

Started in 1939 as a response to the new wartime situation, this Association acted as an umbrella group for feminist groups and individuals to campaign together, under the leadership of Rebecca Sieff. It took over from the International Alliance of Women the publication of *International Women's News* which, until the end of 1945, acted as a British feminist monthly, drawing (as did the Association) on a wide variety of prominent women, particularly Margery Corbett Ashby, and members of the **Six Point Group**. It successfully supported campaigns for such issues as equal compensation for war injuries, and unsuccessfully campaigned for Dorothy E. Evans's Equal Citizenship (Blanket) Bill. It was also very much concerned with the Woman Power question. It had local branches throughout England and Scotland. After the war it was much less active, especially because of Rebecca Sieff's increasing commitment to the state of Israel as distinct from feminist issues. Its original address was 14 Dartmouth Street, London, SW1, but difficulties with the lease meant that its office was later peripatetic, usually being the home of the secretary. It was finally wound up in 1956.

Women's Royal Air Force *see* Women's Auxiliary Air Corps

Women's Royal Army Corps

After the Second World War, it was planned to make the **Auxiliary Territorial Service** into a permanent force. In May 1946, Mary Tyrwhitt was appointed as Director of the ATS to carry out this reform. Three years later, on 1 February 1949, the Women's Royal Army Corps came into existence. It was 'to provide replacements for officers and men in such employment as may be specified by the Army Council from time to time'. The new force had 'Royal' in its title, with the Queen as its Commandant-in-Chief, and Tyrwhitt insisted that its members had the same ranks as men. Instead of khaki, WRACs wore bottle-green uniforms. In 1977, the force was recognised a combatant corps and from 1981, together with other NATO forces, underwent weapons training with small arms. In 1993, the WRAC was absorbed into the Adjutant-General Corps.

(Terry, 1988)

Women's Royal Naval Service

In November 1917, the Admiralty announced its approval for women to take up various shore duties hitherto performed by naval ratings. The

Women's Royal Naval Service was formed under the leadership of Dame Katherine Furse, who was responsible to the Second Sea Lord for its organisation and administration. Known from their initials as Wrens, they were members of the Royal Navy but wore distinctive uniforms. Like their colleagues in the other two Services, the personnel undertook many tasks, including clerical duties, acting as messengers, waitresses in the officers' messes, as well carrying out more technical tasks. Although it was formed at a late date in the war, the WRNS were completely demobilised from the end of 1919. The Service was revived again as the Second World War approached. Dame Vera Laughton Mathews was appointed as Director in April 1939, having served in the WRNS in the First World War. It was the only women's Service to retain its original title. Initially, although only 1,500 were required, more than 15,000 women volunteered as no age limit was stated. Like the **Auxiliary Territorial Service**, the WRNS released men for active service. More than 80 categories of work on shore-based establishments were undertaken including visual signalling, meteorology, clerking and ship, submarine and torpedo maintenance. From 1941, WRNS served overseas. By the end of 1942, the force had risen to 40,000. It was not until February 1990 that WRNS were allowed to serve on board ship. On 1 November 1993, the Service was subsumed in the Royal Navy.

(Drummond, 1960; Mathews, 1948)

Women's Royal Voluntary Service

Begun in 1938 shortly before the outbreak of the Second World War by Lady Reading as the Women's Voluntary Service, and based on local authority areas, the Service proved valuable during the War, carrying out welfare work for the Armed Forces and providing support for the civilian population during the air raids, manning rest centres, providing food and setting up information centres. Its mobile canteens were particularly conspicuous evidence of its activity during the blitz. After the War, its work became oriented more to general social relief. It was granted the title Women's Royal Voluntary Service in 1966. At present it is a community-based, independent organisation which recruits volunteers to assist government departments, local authorities and voluntary bodies in a range of activities. The WRVS takes part in the meals on wheels scheme, works with young families and play groups, undertakes non-medical work in hospitals, provides drivers for rural and social transport and renders assistance when emergencies such as flooding and other natural disasters occur.

Women's Social and Political Union

This organisation, the prime mover of suffrage militancy, was founded in

Manchester in 1903 as a ginger group to agitate for women's suffrage, especially within the Independent Labour Party; founding members included Emmeline Pankhurst, Christabel Pankhurst, Adela Pankhurst, Teresa Billington (later Billington-Greig), Annie Kenney and Elizabeth Wolstenholme Elmy. It soon left the ILP connection behind, and in late 1905 began militant action, with the consequent imprisonment of members.

In 1906 the Union moved to London, and attracted the support of many prominent feminists, especially Charlotte Despard and Emmeline and Frederick Pethick-Lawrence. At this time the *Daily Mail* coined the disparaging term 'suffragette' for the new young militants, a term which was in fact joyfully adopted by them. In 1907, the Union started publishing its own paper *Votes for Women*.

Also in 1907, there was a constitutional crisis within the organisation; despite the objections of the Pankhursts, many prominent members held an annual general meeting according to the Constitution, and elected a new committee under the leadership of Charlotte Despard. At this point Emmeline Pankhurst declared the election null and void. This led to several months' confusion, with the elected members calling themselves the genuine Women's Social and Political Union. The Pankhurst faction responded by calling themselves the National Women's Social and Political Union, with which they persisted (not always consistently) until 1912. In February 1908, the Despard faction adopted the name of the **Women's Freedom League**, thus becoming a separate organisation in name as well as in fact.

From 1908 onwards, the militancy of the Union developed from disrupting meetings to breaking windows, and from mere imprisonment to hunger-striking in prison. Hunger-striking became a major tactic in the second quarter of 1909, and the government response from the third quarter of that year onwards was force-feeding. After truces in 1910 and 1911 in response to the Conciliation Bills being presented to Parliament at this time, militancy resumed when these were sidelined by the government. From 1912–14 the scale of militancy mounted from window-breaking and general damage to property to massive arson and bombing attacks.

In 1912 the Pethick-Lawrences had been expressing doubts about the escalation of militancy, and were therefore expelled, and took with them, as well as the weekly *Votes for Women*, a number of other militants whose perception of militancy did not coincide with that of Emmeline and Christabel Pankhurst. These dissident militants eventually formed the Votes for Women Fellowship, which was later merged with the **United Suffragists**. From October 1912, the organ of the Women's Social and Political Union became the *Suffragette*.

In January 1914, Sylvia Pankhurst's East London Federation of the WSPU, established in 1912, was expelled from the Women's Social and Political Union, and henceforward was known as the East London Federation of Suffragettes; its organ was the *Woman's Dreadnought*. It followed a strong left Socialist line, and in 1917 changed its name to the Workers' Suffrage Federation, an adult suffrage body. The Federation opposed the 1914 war, and engaged in relief work among the poor.

In August 1914, as a result of the outbreak of war, the Women's Social and Political Union abandoned suffrage agitation for strong support for the war, and for women's war work. In 1915 the *Suffragette* changed its name to *Britannia*, and in 1917 the Women's Social and Political Union became the Women's Party, proposing a platform of opposition to Germany, to a negotiated peace, to Bolshevism, to trade unions and to 'shirkers'. In December 1918, it fielded candidates at the general election, of whom the most prominent was Christabel Pankhurst; none was elected, and the Women's Party does not appear to have survived into 1919.

By no means all members of the Women's Social and Political Union supported its war policy. Two breakaway organisations were formed to maintain the campaign for the vote: the Suffragettes of the WSPU in 1915, and the Independent Women's Social and Political Union in 1916. Neither of these appears to have lasted beyond 1918.

(Holton, 1986; Pankhurst, 1931; Rosen, 1974)

Women's Tax Resistance League

This began life as an offshoot of the **Women's Freedom League**, and was formally established by Margaret Kineton Parkes in 1909. Its object was to further the cause of women's suffrage by refusing to pay taxes. Members regularly did this, and had their goods distrained and auctioned with much publicity, though they were often bought back by friends. In some cases, resistance was more protracted, resulting in house sieges by bailiffs. In theory, as a means of non-violent resistance, it drew on all suffrage tendencies, but the majority of members tended to be from the Women's Freedom League and **Women's Social and Political Union**. It effectively ceased activity after the outbreak of war in 1914.

(Holton, 1996; Joannou and Purvis, 1998; Parkes, 1919)

Women's Timber Corps

Established as a special section of the **Women's Land Army** in April 1942 and administered by the Home Timber Production Department of the Ministry of Supply. The uniform was the same as for the WLA except that members wore a special green beret rather than a hat and the badge displayed a fir tree instead of a wheat-sheaf. Its tasks included the felling and crosscutting of trees, clearing woods and driving tractors and lorries. Before it became a special section, there were already 1,000 WLA members working for the Department and 250 volunteers a month were attracted after it was set up. Members had to be over 21 and training at special training centres lasted a month, where participants were introduced to a range of occupations. The hours were 7 a.m. to 4.30 p.m. with a half-day off on Saturdays. During training, recruits were given a 10 shillings a week allowance and at the end of training, 45 shillings a week. Recruitment was terminated in July 1943 as other claims on women power were considered more urgent. At this date, 3,900 women in England and Wales were member of the Corps and 1,000 in Scotland. The Women's Timber Corps was disbanded in 1946.

(*Meet the Members*, 1945)

Women's Total Abstinence Union *see* **British Women's Temperance Association**

Women's Trade Union Association

This Association was founded in 1889 by Clementina Black, formerly secretary of the **Women's Trade Union League**. Black concluded that low wages were at the root of working women's problems and that this could be overcome by founding unions. Black also argued that employers should fix selling prices for goods among themselves and agree to pay minimum union rates rather than indulge in wasteful competition. She also recommended that a Consumers' League should be established to visit places of employment and publish their findings so that the public would be able to judge who were good employers. Later, believing that women's trade unions were not certain of their future, the Association became a political pressure group and joined the **Women's Industrial Council**.

(Mappen, 1986)

Women's Trade Union League

The League, which was founded in 1874, had four main objectives: to protect the trade interests of its members by endeavouring to prevent the undue depression of wages and to campaign for equal hours of work with men; to provide a fund for members who were unemployed or ill; to collect and disseminate useful information on employment opportunities; and to encourage arbitration in cases of dispute between employer and employee. The originator of the League was Emma Paterson, who had served a brief apprenticeship as a bookbinder. The year before, in 1873, she had married and with her husband visited America. On a visit to New

York, they were impressed to note that some of the most successful unions were managed by working women; on their return to England she created public interest by publishing articles on the subject.

Called at first the Women's Protective and Provident League, it soon changed its title to the Women's Trade Union League. Although it was essentially established to promote trade unionism among women, it was not hostile to employers and among its members were several men, including Charles Kingsley and Arnold Toynbee. Mrs Paterson, as honorary secretary, threw her energies into organising a **Society of Women Employed in Bookbinding**. This was followed by a Society of Dressmakers, Milliners and Mantlemakers, a Society of Women Employed in Binding, Sewing and Trimming Men's Hats, a Society of Upholstresses and a Society of Shirt and Collar Makers. All this was achieved within the first year. These Societies met with varying fortunes, some incurring strong male opposition or failing through lack of funds. In some parts of the country, the League flourished, attracting especially textile workers. For example, at Dewsbury, Yorkshire, over 1,000 women joined a woollen weavers' union which conducted an eight-weeks' strike following a wage reduction and resulting in a successful outcome. The different Societies of the League sent delegates to the Trades Union Congress and read papers on women's questions. From 1877, the League invited delegates to a women's conference during the annual Congress week. Eighty members attended the first meeting and this became a regular feature of future Congresses.

After the 1878 Factory and Workshops Bill became law, women delegates to the Congress, in the face of opposition from the men, proposed that women factory inspectors should be appointed; it was not until 1892 that the Congress accepted this resolution. Suspicion from men that employers, by employing women, would lead to undercutting of wages, was matched by the League's determination not to be excluded from occupations. Thus, in 1882, the League opposed a Bill which would have prohibited young girls under the age of 14 from acting as blacksmiths on the grounds that legislation should state that it applied to both boys and girls alike. Between 1874 and 1884, the League was able to establish between 30 and 40 Societies in England and Scotland. Emma Paterson's early death occurred in 1886. In 1903 Margaret Bondfield, a member of the League's Committee, recommended her friend, Mary Macarthur, then aged 23, as secretary. Lady Emilia Dilke, the first president of the League, aimed at admitting women as full members of men's unions. By 1914, some 350,000 women were members of trades unions. It became part of the Trades Union Congress in 1921.

(Drake, 1984 [1920])

Women's True Temperance Committee

Apparently founded by Lady Astor in about 1925, it published a periodical called variously *Woman's View* or *Women's View*, devoted to the cause of abstinence from alcohol. It does not appear to have survived after 1929.

Women's Unionist and Tariff Reform Association

This Association was closely connected with National Union of Conservative Associations but was constitutionally separate. It transferred to the Central Office as the Women's Unionist Association in 1928. One of its main tasks subsequently was to train women as organisers as well as promoting lectures and visiting constituents. It had direct representation on the National Union's Council.

Women's University Settlement

In the 1880s university settlements of young men living among the poor to help them were being established, starting with Toynbee Hall in the East End of London in 1884. In 1887, the Cambridge Ladies' Discussion Society, after a talk by Henrietta Barnett, determined to establish a similar venture for women. Representatives of Newnham and Girton Colleges at Cambridge, and Somerville and Lady Margaret Hall at Oxford came together to form the Women's University Association, based at Alice Grüner's house, 44 Nelson Square, Southwark, with Alice Grüner as Head Worker. Initially the Council consisted of two representatives each from Newnham and Girton, one each from Somerville and Lady Margaret Hall, and the Head Worker. The Settlement opened its doors in September 1887; there was immediate co-operation with the Charity Organisation Society, the London Pupil Teachers' Association, the **Metropolitan Association for Befriending Young Servants** and the Children's Country Holiday Fund. The first annual general meeting took place in the Blackfriars Road Temperance Hall in 1888 and was chaired by Mary Paley Marshall. In 1889 the Association's name was changed to the Women's University Settlement; in that year Miss Grüner resigned as Head Worker, and her place was eventually taken, this time as paid Warden, by Margaret Sewell in 1891. By 1895 the Settlement had 31 resident workers, and 61 non-resident. It organised holiday treats for local children, and arranged special music and dancing classes as well as a debating society and reading parties. Its activities expanded during the year up to the First World War, and included the formation of children's clubs, especially the Acland Club for boys and girls, and in 1899 the St Mary's Club for Working Girls at Union Street in the Borough, with St Monica's Home of Rest at Birchington-on-Sea, Kent. It was also involved in St Crispin's

Workshop, a cripples' bootmaking shop on Southwark Bridge Road, and by the time that Helen Gladstone became Warden in 1901 it was closely involved in social work of various sorts with Southwark Council.

In 1921, it acquired 47 Nelson Square, where it built a hall. By the outbreak of the Second World War it had also initiated a baby clinic, a Girl Guides troop and an unemployed men's club. In 1940–41 the premises suffered bomb damage, and this, with uncertainties over post-war planning, made the Settlement's situation extremely precarious. In 1958, faced with a major financial crisis, it had to terminate the employment of paid staff and close the residential part. Subsequently it transformed itself into something closer to a community centre, and in 1961 it adopted a new constitution and changed its name to the Blackfriars Settlement.

(Barrett, 1985)

Women's Vegetarian Union

Founded in 1895 by Mme Alexandrine Veigelé, its aim was to promote vegetarianism among women. Its first treasurer was Miss E. Wardlaw Best, who was succeeded by Mrs T. R. Allinson; the honorary secretary was Mlle Adrienne Veigelé. It started at 96 Crawford Street, Baker Street, London, W, then moved first to 87 Praed Street, then to 37 Hereford Road, Bayswater, London, W. It operated an agency for domestic servants which guaranteed staff who were vegetarians, or willing to adopt vegetarianism. The minimum annual subscription was 2s 6d; in 1896 membership was over 200. It was still active in 1901.

Women's Voluntary Service *see* Women's Royal Voluntary Service

Women's Volunteer Reserve

Shortly after the outbreak of the First World War, the Women's Volunteer Reserve was formed by Lady Edith Londonderry, Lady Tichfield, Lady Pembroke and Edith Montgomery. Volunteers, who wore khaki uniforms and khaki felt hats to emphasise their military status, underwent drill and carried out a variety of unattractive jobs, including digging allotments, scrubbing floors, and helping with the salvage of waste paper. Lady Londonderry, as Colonel-in-Chief, broke away from the Reserve in 1915 and formed the **Women's Legion**.

Work Girls' Protection Society

Founded by the Sisters of St Mary's, an Anglican Order, the Society flourished in the 1880s. It aimed at raising the position of machinist girls by encouraging them to go into high-class domestic service. A School of

Instruction was established to enable them to make a living and an Evening Club was provided for recreation and amusement. The Society opened a shelter for the respectable homeless, gave relief in food and clothing and arranged for the products of poor workhouse women to be sold at depots. In 1889, some 5,004 girls and women had benefited from the working of the Society.

(Church of England, 1889)

Workers' Birth Control Group

Formed in 1924, its leading members included Mrs L'Estrange Malone, Dora Russell, Susan Lawrence, Dorothy Jewson, Ernest Thurtle, H. G. Wells and Frida Laski. Its purpose was to campaign for birth control within the Labour movement and specifically the Labour Party. Its object was to emphasise birth control as a maternal health measure, as distinct from the frequently class-based eugenic arguments of the Malthusian League and other bodies unlikely to find favour with the political left. It was vigorous in campaigning against the Ministry of Health's ban on the giving of birth-control information in Maternity Centres. In 1930 it affiliated with the **National Birth Control Association**, and dissolved itself as a separate entity in 1931.

(Leathard, 1980)

Workers' Suffrage Federation *see* Women's Social and Political Union

Workhouse Visiting Society

Founded in 1858 by Louisa Twining following her private visits to workhouses and reporting on their conditions. The Society's aim was 'to promote the moral and spiritual improvement of workhouse inmates … and to provide a centre of communication for all persons interested in that object'. Initially, both Guardians and the Poor Law Board rejected the offer; the Board later grudgingly recognised the Society whilst many Guardians were able to ignore it. It began as a sub-committee of the National Association for the Promotion of Social Science and within a year there were 340 members who visited London and provincial workhouses. Many abuses were uncovered and the Society attempted to ameliorate the lives of the inmates by effecting minor improvement during the members' visits. In 1861, Baroness Burdett-Coutts gave funds for a home to train workhouse girls as servants and helped to find employment for unmarried women. Although the Society was disbanded in 1865, it led to demand for women to stand as Poor Law Guardians.

(Crowther, 1982)

Working Ladies' Guild

It was formed in 1877 by the Lady Mary Fielding with the object of providing employment for, and otherwise assisting, 'gentlewomen whose means are small, but whose education and training unfit them for the rude competition now necessary for any chance of success'. The fundamental principle of the Guild was that each Associate should give time, influence or money (if possible, all three). Every lady entered on the books of the Guild for assistance had to be personally well known to her introducing Associate, who then became her guarantor. The first meeting consisted of 12 ladies, but by 1888 this had risen to more than 1,300 Associates. In 1887 nearly £3,000 had been raised, of which £1,700 was spent on grants; 300 ladies had been helped in other ways. The Guild's president was the Archbishop of Canterbury, with HRH Princess Beatrice as president of Departments (Art and Needlework, Painting, Plain Work, Fancy Work), and HRH Princess Frederica as President of Groups. The Guild's offices were at 217 Brompton Road, London, SW, and the secretary's address was 16 Portman Square, London. It was still active in 1909.

Working Women's Club *see* **Women's Club**

World Women's Committee Against War and Fascism (British Section)

This began in 1936 as a Popular Front organisation, theoretically with no sectarian or party links, but in fact closely associated with the Communist Party. Among its members were strong Labour supporters such as Selina Cooper, Jennie Lee and Ellen Wilkinson as well as middle-class feminists including Vera Brittain and Maud Royden, and left-leaning old suffragists, such as Emmeline Pethick-Lawrence and Charlotte Despard. However, the Labour Party expelled individuals for joining. Its organ was *Woman Today*, edited by Charlotte Haldane, which ran from 1936–40, being revived from 1944–48 as an openly Communist periodical edited by Tamara Rust.

(Bruley, 1985)

Writers' Club

Founded in 1892 by the journalist Frances Low at 10 Norfolk Street, near Fleet Street, London, it claimed to be unique in being the only Club devoted to women of one profession. Entry, which was limited to 300, was based on evidence of literary or journalistic work. Entrance fee was one guinea for town members and the same amount for the annual subscription. Many well-known authors were members and a quiet room was reserved for writing. 'At Homes' were held every Friday afternoon when guests could be invited to tea. No residential accommodation was provided and silence was enforced in the Writing Room. In the early 1900s a group of members, dissatisfied with the Club's lack of physical amenities, broke away under the leadership of Constance Smedley, to form the **Lyceum Club**. The Writers' Club was still in existence in the 1920s.

Yorkshire Ladies' Council of Education

The Council, formed in 1886, started its existence as a committee of the West Riding Board of Education to organise the Cambridge Local Examination for Girls. When the West Riding Board was made responsible in 1870 for a wider area, changing its name to the Yorkshire Board of Education, the ladies' committee formally became the Ladies' Honorary Council of the Board. The latter body pursued the object of providing general elementary education for working girls and women, and encouraging the provision of suitable higher education. A separate group of ladies had formed the Leeds Ladies' Educational Association in 1867. They were given the task of organising the Cambridge Local Examinations which led to some friction between the two parties. In 1875, the Ladies' Honorary Council separated from the Yorkshire Board of Education and four years later took over the Registry of Governesses originated by the Leeds Ladies' Educational Association. By the early 1880s, the two bodies had merged.

(Barrow, 1981)

Young Women's Christian Association

In 1835, Mrs Emma Robarts set up a prayer group for women to pray for each other and for young women. She also established a home for 'that large class of young women employed in business, destitute of friendship'. As a result, the YWCA was founded by the Hon. Mrs (later Lady) Kinnaird in 1861 to help the many young women who were being attracted to London and other large cities by employment possibilities. The main concern of the Association was to make suitable arrangements for the welfare of this vulnerable group by providing boarding houses, institutes, clubs, libraries, restaurants and holiday homes for their use. The first hostel was opened in Upper Charlotte Street, Fitzroy Square, London, where young women above the rank of domestic servants were lodged for a guinea a week. The aims of the Association were to unite young women for prayer and to promote their moral, social and intellectual well-being. The title of Young Women's Christian Association was first used in 1876. By 1893, there were branches in many parts of Great Britain – one had been opened in Dublin in 1866 – and two divisions catered for Europe and the Colonies. Branches were formed in many other parts of the world, including Japan and the USA. At this time, there were as many as 100,000 members world-wide.

(Harrison, 1973)

Young Women's Help Society

This Society found work for poor ladies and had alleviated distress among Irish ladies whose income depended on rent increases. At its 1886 annual general meeting, the Society reported that some 4,000 girls had benefited during the year from its activities and that 19 new branches had been formed, bringing the total to 72.

BIBLIOGRAPHY

A. J. R., *Suffrage Annual and Women's Who's Who,* Stanley Paul, 1913.

Abel-Smith, Brian, *A History of the Nursing Profession*, Heinemann, 1960.

Alexander, Sally (ed.), *Women's Fabian Tracts* (Women's Source Library), Routledge, 1987.

All-England Women's Hockey Association, *Women's Hockey: From Village Green to Wembley*, Macdonald & Evans, 1950.

Allan, Peter and Allan, Moya (eds), *Nursing, Midwifery and Health Visiting since 1900*, Faber, 1982.

Andrews, Maggie, *The Acceptable Face of Feminism*, Lawrence & Wishart, 1997.

Angell, Shirley, *Pinnacle Club: A History of Women Climbing*, The Pinnacle Club, 1988.

Bailey, Jack, *The British Co-operative Movement*, Hutchinson, 1955.

Baly, Monica E., *A History of the Queen's Nursing Institute*, The Institute, 1987.

Baly, Monica E., *Nursing and Social Change*, 3rd edn, Routledge, 1995.

Barrett, Gladys, *Blackfriars Settlement: A Short History 1887–1987*, The Settlement, s.d.

Barrow, Margaret, *Women 1870–1928: A Select Guide to Printed and Archival Sources in the United Kingdom*, Mansell Publishing, 1981.

Beddoe, Deirdre, *Back to Home and Duty*, Pandora, 1989.

Bell, Enid H. C. Moberly, *Storming the Citadel: The Rise of the Woman Doctor*, Constable, 1953.

Bell, Enid H. C. Moberly, *The Story of Hospital Almoners: The Birth of a Profession*, Faber, 1961.

Bew, Paul, *Land and the National Question in Ireland, 1858–82*, Humanities Press, 1979.

Black, Eugene C., *The Social Politics of Anglo-Jewry, 1880–1920*, Blackwell, 1988.

Bolt, Christine, *The Women's Movements in the United States and Britain from the 1790s to the 1970s*, Harvester Wheatsheaf, 1993.

Boston, Sarah, *Women Workers and the Trade Union Movement*, Davis-Poynter, 1980.

Boston, Sarah, *Women Workers and the Trade Unions*, revised edn, Lawrence & Wishart, 1987.

Bowman, Gerald, *The Lamp and the Book: The Story of the Royal College of Nursing, 1916–1946,* Queen Anne Press, 1967.

Bowser, Thekla, *The Story of British VAD Work in the Great War*, Melrose, 1917.

Braine, Sheila E., 'London's Clubs for Women', in G. R. Sims (ed.), *Living London*, Vol. 1, Section 1, Cassell, 1902.

Brion, Marion, *Women in the Housing Service*, Routledge, 1995.

British Federation of University Women, *The History of the British Federation of University Women, 1907–1957*, The Federation, 1957.

Bruley, Sue, 'Women Against War…', in J. Fyrth (ed.), *Britain, Fascism and the Popular Front*, Lawrence & Wishart, 1985.

Bundock, Clement J., *The National Union of Printing, Binding and Paper Workers,* Oxford University Press, 1959.

Bunt, Sidney, *Jewish Youth Work in Britain: Past, Present and Future*, Bedford Square Press, 1975.

Burdett-Coutts, Angela Georgina (ed.), *Women's Mission: A Series of Congress Papers on the Philanthropic Work of Women*, Sampson Low, Marston & Co., 1893.

Burman, Rickie, 'Middle-Class Anglo-Jewish Lady Philanthropists and Eastern European Jewish Women', in Joan Grant (ed.), *Women, Migration and Empire*, Trentham Books, 1996.

Burstyn, Joan N., *Victorian Education and the Ideal of Womanhood,* Croom Helm, 1980.

Bush, Julia, *Edwardian Ladies and Imperial Power*, Leicester University Press, 2000.

Church of England, *Official Year Book of the Church of England*, SPCK, 1889, 1906.

Colclough, M., *Women's Legion, 1916–1920*, Spearman, 1920.

Cooke, Douglas (ed.), *Youth Organisations of Great Britain*, Jordan & Sons, 1946.

Copelman, Diana M., *London Women Teachers: Gender, Class and Feminism, 1870–1930*, Routledge, 1996.

Cossey, Rosalynde, *Golfing Ladies: Five Centuries of Golf in Great Britain and Ireland*, Orbisa, 1984.

Cowper, Julia Margaret, *The Auxiliary Territorial Service: The Second World War, 1939–1945,* War Office, 1949.

Crawford, Elizabeth, *The Women's Suffrage Movement: A Reference Guide,* UCL Press, 1999.

Crow, Duncan, *The Victorian Woman*, Allen & Unwin, 1971.

Crowther, M.A., *The Workhouse System, 1834–1929: The History of an English Social Institution*, University of Georgia Press, 1982.

Daunt, Marjorie, *By Peaceful Means: The Story of Time and Talents, 1887–1987*, Time and Talents Association, 1989.

Davidson, Caroline, *A Woman's Work is Never Done: A History of Housework in the British Isles, 1650-1950*, Chatto & Windus, 1982.

De Courcy, Anne, *Circe: The Life of Edith, Marchioness of Londonderry*, Sinclair-Stevenson, 1992.

Deakin, Phyllis, *Press On,* H. E. Walter, 1984.

Deepwell, Katy, 'A History of the Society of Women Artists', in C. B. De Laperrière (ed.), *The Society of Women Artist Exhibitors, 1855–1996,* Hilmarton Manor Press, 1996.

Dodd, Christopher, 'Rowing', in Tony Mason (ed.), *Sport in Britain: A Social History*, Cambridge University Press, 1989.

Doughan, David and Sanchez, Denise, *Feminist Periodicals, 1855–1984*, Harvester Wheatsheaf, 1987.

Dowling, W. C., 'The Ladies' Sanitary Association and the Origin of the Health Visitor Service', London University, MA thesis, 1963.

Drake, Barbara, *Women in Trade Unions*, Virago,1984 (first published 1920).

Drummond, John D., *Blue for a Girl: The Story of the WRNS*, W. H. Allen, 1960.

Dyhouse, Carol, *Feminism and the Family in England, 1880–1938*, Blackwell, 1989.

Dyhouse, Carol, *Girls Growing Up in Late Victorian and Edwardian England*, Routledge & Kegan Paul, 1981.

Ellsworth, Edward W., *Liberators of the Female Mind: The Shirreff Sisters, Educational Reform and the Women's Movement*, Greenwood, 1979.

Englishwoman's Review, 1878–1908.

Englishwoman's Year Book, 1912.

Eoff, Shirley, *Viscountess Rhondda, Equalitarian Feminist*, Ohio State University Press, 1991.

Escott, Beryl E., *Women in Air Force Blue*, Patrick Stephens, 1989.

Faulkner, Madge, *The British Federation and Business and Professional Women,* The Federation, 1969.

Foster, R. F., *Charles Stuart Parnell: The Man and His Family*, Harvester, 1976.

Franz, Nellie Alden, *English Women Enter the Professions* (privately printed), 1965.

Fredeman, William E., 'Emily Faithfull and the Victoria Press', *The Library* (June 1973).

Freeman, Flora Lucy, *Religious and Social Work amongst Girls*, Skeffington, 1901.

Furse, Katherine, *Hearts and Pomegranates: The Story of Forty-five Years, 1875 to 1920*, Peter Davies, 1940.

Gaffin, Jean and Thoms, David, *Caring and Sharing: The Centenary History of the Co-operative Women's Guild*, Co-operative Union, 1983.

Gaze, Delia (ed.), *Dictionary of Women Artists*, Vol. I, Fitzroy Dearborn, 1997.

Gleadle, Kathryn, *The Early Feminists*, Macmillan, 1995.

Glenday, N. and Price, N., *Reluctant Revolutionaries: A Century of Headmistresses, 1874–1974*, Pitman, 1974.

Glick, D., *The National Council of Women of Great Britain: The First One Hundred Years, 1895–1994*, National Council of Women of Great Britain, 1995.

Goldsmith, Margaret, *Women at War*, Lindsay Drummond, 1943.

Gordon, Peter (ed.), *Politics and Society: The Journals of Lady Knightley of Fawsley, 1885 to 1913*, Northamptonshire Record Society, 1999.

Gosden, P. H. J. H., *The Evolution of a Profession: A Study of the Contribution of Teachers' Associations to the Development of School Teaching as a Professional Occupation*, Blackwell, 1972.

Governess, 'The Metropolitan Board Mistresses' Association', June 1882.

Grant, Joan (ed.), *Women, Migration and Empire*, Trentham Books, 1996.

Gwynne-Vaughan, Helen, *Service with the Army*, Hutchinson, 1942.

Hall, Dorothy, *Making Things Happen*, National Federation of Business and Professional Women's Clubs of Great Britain and Northern Ireland, 1963.

Hallsworth, Joseph and Davies, R. J., *The Working Life of Shop Assistants: A Study of Conditions of Labour in the Distributive Trades*, National Labour Press, 1910.

Hammerton, James A., *Emigrant Gentlewomen*, Croom Helm, 1979.

Harrison, Brian, 'For Church, Queen and Family: The Girls' Friendly Society, 1874–1920', *Past & Present*, 61 (November 1973).

Harrison, Brian, *Prudent Revolutionaries*, Clarendon Press, 1987.

Harrison, Brian, *Separate Spheres*, Croom Helm, 1978.

Hart, Christopher, *Behind the Mask. Nurses: Their Unions and Nursing Policy*, Baillière Tindall, 1994.

Heath-Stubbs, Mary, *Friendship's Highway: Being the History of the Girls' Friendly Society, 1875–1935*, Girls' Friendly Society, 1935.

Hinton, James, 'Militant Housewives: The British Housewives' League and the Attlee Government', *History Workshop Journal*, 38 (Autumn 1994).

The History of the British Federation of University Women, 1907–1957, The Federation, 1957.

Holcombe, Lee, *Victorian Ladies at Work*, David & Charles, 1973.

Holcombe, Lee, *Wives and Property*, Martin Robertson, 1983.

Holledge, Julie, *Innocent Flowers*, Virago, 1981.

Hollis, Patricia, *Ladies Elect: Women in English Local Government, 1865–1914*, Clarendon Press, 1987.

Hollis, Patricia, 'Women in Council: Separate Spheres, Public Space', in Jane Rendall (ed.), *Equal or Different: Women's Politics, 1800–1914*,

Blackwell, 1986.
Holton, Sandra Stanley, *Feminism and Democracy, 1900-1918*, Cambridge University Press, 1986.
Holton, Sandra Stanley, *Suffrage Days*, Routledge, 1996.
Hopkins, Jane Ellice, *Ladies' Associations for the Care of Friendless Girls* (reprinted from *The Day of Rest*, 1878).
Horn, Pamela, 'Training and Status in Domestic Service: The Role of the League of Skilled Housecraft, 1922–42', *History of Education Society Bulletin*, 65 (May 2000).
Hume, Leslie Parker, *The National Union of Women's Suffrage Societies, 1897–1914*, Garland, 1982.
Humphreys, B.V., *Clerical Unions in the Civil Service*, Blackwell & Mott, 1958.
Jameson, Anna Brownell, *Sisters of Charity, Catholic and Protestant, Abroad and at Home* [s.n.], 1855.
Jerman, Betty, *The Lively-Minded Women,* Heinemann, 1981.
Jex-Blake, Sophia, 'Medical Women', *Nineteenth Century*, November (1887).
Joannou, Maroula and Purvis, June, *Women's Suffrage: New Feminist Perspectives*, UCL Press, 1998.
John, Angela and Eustance, Claire (eds), *The Men's Share?*, Routledge, 1997.
Joy, Nancy, *Maiden Over*, Sporting Handbooks, 1950.
Judge, Roy, 'Mary Neal and the Espérance Morris', *Folk Music Journal*, 5, 5 (1989).
Kadish, Sharman, *'A Good Jew and a Good Englishman': The Jewish Lads' and Girls' Brigade, 1895–1995*, Vallentine Mitchell, 1996.
Kamm, Josephine, *Hope Deferred: Girls' Education in English History*, Methuen, 1965.
King, Peter, *Women Rule the Plot: The Story of the 100 Year Fight to Establish Women's Place in Farm and Garden*, Duckworth, 1999.
Law, Cheryl, *Suffrage and Power,* I. B. Tauris, 1997.
Lawton, Denis and Gordon, Peter, *HMI*, Routledge, 1987.
Leathard, Audrey, *The Fight for Family Planning*, Macmillan, 1980.
Leeds, H., *The Origin and Growth of the Union of Practical Suffragists*, The Union, 1898.
Lehmann, Liza, *Liza Lehmann*, T. Fisher Unwin, 1919.
Lemoine, Sheila C., 'The North of England Council for Promoting the Higher Education of Women', University of Manchester, MEd thesis, 1968.
Leneman, Leah, *A Guid Cause*, Mercat Press, 1995.
Levenson, Leah and Natterstad, Jerry H., *Hannah Sheehy-Skeffington: Irish Feminist*, Syracuse University Press, 1986.
Lewenhak, Sheila, *Women and Trade Unions*, Benn, 1977.

Lewis, Gifford, *Eva Gore-Booth and Esther Roper*, Pandora, 1988.
Lewis, Jane, *Women in England, 1870–1950: Sexual Divisions and Social Change*, Wheatsheaf Books, 1984.
Liddington, Jill and Norris, Jill, *One Hand Tied Behind Us*, Virago, 1978.
Linklater, Andro, *An Unhusbanded Life*, Hutchinson, 1980.
Lochhead, M., *A Lamp Was Lit: Girls' Guildry through Fifty Years*, Moray Press, 1949.
Lock, Joan, *The British Policewoman*, Robert Hale, 1979.
Lovenduski, J., Norris, P. and Burness, C., ' The Party and Women', in A. Seldon and S. Ball (eds), *The Conservative Century: The Conservative Party since 1900*, Oxford University Press, 1994.
Lowndes, G.A.N., *Silent Social Revolution*, Oxford University Press, 1950.
Luddy, Maria, *Women and Philanthropy in Nineteenth-Century Ireland*, Cambridge University Press, 1995.
McCleary, G. F. M., *The Maternity and Child Welfare Movement*, P. S. King, 1935.
McCrone, Kathleen E., *Sport and the Physical Emancipation of Englishwomen, 1870–1914*, Routledge, 1988.
McFeely, Mary Drake, *Lady Inspectors: The Campaign for a Better Workplace, 1893–1921*, Blackwell, 1988.
McGann, Susan, *The Battle of the Nurses*, Scutari Press, 1992.
McHugh, Paul, *Prostitution and Victorian Social Reform,* Croom Helm, 1980.
McIntosh, Peter C., *Physical Education in England since 1800*, Bell & Hayman, 1968.
Macnicol, John, *The Movement for Family Allowances, 1918–45: A Study in Social Development Policy*, Heinemann, 1980.
Magnus, L., *The Jubilee Book of the Girls' Public Day School Trust, 1873–1923*, Cambridge University Press, 1923.
Mappen, Ellen F., 'Strategists for Change: Social Feminist Approaches to the Problems of Women's Work', in Angela V. John (ed.), *Unequal Opportunities: Women's Employment in England, 1800–1918*, Blackwell, 1986.
Mason, Francis, 'The Newer Eve: The Catholic Women's Suffrage Society in England, 1911–1923', *Catholic Historical Review*, 52, 4 (October 1986).
Mason, Tony (ed.), *Sport in Britain: A Social History*, Cambridge University Press, 1989.
Mathews,Vera Laughton, *Blue Tapestry*, Hollins & Carter, 1948.
Matthews, J.J., 'They Had Such a Lot of Fun', *History Workshop Journal*, 30 (Autumn 1990).
Mayhall, Laura E., 'Creating the Suffragette Spirit', *Women's History Review*, 4, 3 (1995).

Meet the Members: A Record of the Timber Corps of the Women's Land Army, Bennett Brothers, 1945 (reprinted Imperial War Museum, 1997).

Merz, Caroline, *After the Vote: The Story of the National Union of Townswomen's Guilds in the Year of its Diamond Jubilee*, The Union, 1988.

Messenger, Rosalind, *The Doors of Opportunity*, Femina, 1967.

Middleton, Lucy (ed.), *Women in the Labour Movement: The British Experience*, Croom Helm, 1977.

Monk, Una, *New Horizons*, HMSO, 1963.

Nursing in the Army: Queen Alexandra's Imperial Nursing Service, HMSO, 1905.

Official Year-Book of the Church of England, SPCK, 1906.

Owen, David, *English Philanthropy, 1660–1960*, Oxford University Press, 1965.

Paine, W., *Shop Slavery and Emancipation: A Revolutionary Appeal to the Educated Young Men of the Middle Class*, P. S. King, 1912.

Pankhurst, Estelle Sylvia, *The Suffragette Movement*, Longman, 1931 (reprinted Virago, 1978).

Parkes, Margaret Kineton, *History of Women's Tax Resistance*, Women's Tax Resistance League (*c*. 1919).

Parnell, Nancy Stewart, *A Venture in Faith*, St Joan's Alliance, 1961.

Pavey, Angel E., *The Story of Nursing as an Art, a Vocation, and a Profession*, 3rd edn, Faber, 1951 (first published 1938).

Philipps, M. Wynford *et al.*, *A Dictionary of Employments Open to Women*, Women's Institute, 1898.

Phipps, E., *History of the National Union of Women Teachers,* The Union, 1928.

Popham, Hugh, *F.A.N.Y. The Story of the Women's Transport Service*, Leo Cooper/Secker & Warburg, 1984.

Pratt, Edwin A., *A Woman's Work for Women: Being the Aims, Efforts and Aspirations of Miss Louisa M. Hubbard*, Newnes, 1898.

Prochaska, Frank K., *Women and Philanthropy in Nineteenth Century England*, Clarendon Press, 1980.

Pugh, Martin, *Women and the Women's Movement in Britain, 1914–1959*, Macmillan, 1992.

Rae, Lettice Milne (ed.), *Ladies in Debate: Being a History of the Ladies' Edinburgh Debating Society, 1865–1935*, Oliver & Boyd, 1936 .

Ranyard, Ellen, *Nurses for the Needy or Bible-Women Nurses in the Homes of the London Poor* [s.n.], 1875.

Robb, J. H., *The Primrose League*, AMS Press, 1942.

Rosen, Andrew, *Rise Up, Women!*, Routledge, 1974.

Rover, Constance, *Women's Suffrage and Party Politics, 1866–1914*, Routledge, 1967.

Rubinstein, David, *Before the Suffragettes: Women's Emancipation in the 1890s*, Harvester, 1986.

Rubinstein, David, *A Different World for Women: The Life of Millicent Garrett Fawcett*, Harvester Wheatsheaf, 1991.

Sherr, Rosalie, *The Story of the Federation of Women Zionists of Great Britain and Ireland*, The Federation, 1968.

Shiman, Lilian Lewis, *Women and Leadership in Nineteenth Century England*, Macmillan, 1992.

Shreir, Sally (ed.), *Women's Movements of the World*, Longman, 1988.

Sillitoe, Helen, *A History of the Teaching of Domestic Subjects*, Methuen, 1933.

Sims, G. R., *Living London*, Vol. I, Section 1, Cassell, 1902.

Smedley, Constance, *Crusaders*, Duckworth, 1929.

Smith, Harold L., 'The Politics of Conservative Reform', *Historical Journal*, 35, 2 (1992).

Smith, Harold L., *The British Women's Suffrage Campaign, 1866–1928* (Seminar Studies in History), Addison-Wesley Longman, 1998.

Spender, Dale, *Women of Ideas: And What Men Have Done to Them*, Routledge, 1983.

Springhall, John, *Youth, Empire and Society: British Youth Movements, 1883–1940*, Croom Helm, 1977.

Springhall, John, Fraser, Brian and Hoare, Michael (eds), *Sure and Steadfast: A History of the Boys' Brigade, 1883 to 1983*, Collins, 1983.

Stevenson, Janet, *A History of the Ladies Guild of Change Ringers*, The Guild, 1999.

Stocks, Mary, *A Hundred Years of District Nursing*, Allen & Unwin, 1960.

Stowell, Sheila, *A Stage of Their Own*, Manchester University Press, 1992.

Strachey, Ray, *The Cause: A Short History of the Women's Movement*, Bell, 1928 (reprinted Virago, 1978).

Strachey, Ray, *Women's Suffrage and Women's Service*, London and National Society for Women's Service, 1927.

Tancred, Edith, *Women Police*, National Council of Women of Great Britain, 1951.

Taylor, Barbara, *Eve and the New Jerusalem*, Virago, 1983.

Taylor, Eric, *Women Who Went to War, 1938–40*, Grafton Books, 1988.

Terry, R., *Women in Khaki*, Columbus Books, 1988.

Thompson, Vronwyn M., *1910 ... And Then? A Brief History of the Girl Guides Association*, The Guide Association, 1990.

Tickner, Lisa, *The Spectacle of Women*, Chatto, 1987.

Tidcombe, Marianne, *Women Bookbinders, 1880–1920*, Oak Knoll Press and the British Library, 1996.

Tooley, Sarah A., 'The Women's Liberal Federation', *Lady's Realm*, 1897.

Tropp, A., *The School Teachers*, Heinemann, 1957.

Twining, Louisa, *Workhouses and Pauperism*, Methuen, 1898.

Tyrer, Nicola, *They Fought in the Fields: The Women's Land Army: The Story of a Forgotten Victory*, Sinclair-Stevenson, 1996.

VanArsdel, Rosemary T., *Mrs Florence Fenwick-Miller, Feminism and the Woman's Signal*, University of Puget Sound, 1979.

Wadge, D. Corbett, *Women in Uniform*, Sampson, Low, Marston, 1946.

Walkowitz, Judith R., *Prostitution and Victorian Society*, Cambridge University Press, 1980 .

Ward, Audrey, 'Physiotherapy', in Rosalind Silverstone and Audrey Ward (eds), *Careers of Professional Women*, Croom Helm, 1980.

Webb, Catherine, *The Woman with the Basket: The History of the Women's Co-operative Guild, 1883–1927*, Manchester Co-operative Wholesale Society, 1927.

Webster, F. R. M., *Athletics of Today for Women: History, Development and Training*, Frederick Warne, 1930.

Whitelaw, Lis, *The Life and Rebellious Times of Cicely Hamilton*, Women's Press, 1990.

Williams, Jack, 'Cricket', in Tony Mason (ed.), *Sport in Britain: A Social History*, Cambridge University Press, 1989.

Wilmott, Phyllis, 'The Part-time Social Worker', *Case Conference*, 7, 11 (May 1961).

Women's National Commission, *Women's Organisations in the United Kingdom, 1997–98*, The Commission, 1997.

The Women's Printing Society, The Society, 1913.

Women's Studies International Forum, 10, 3 (1987) (whole issue).

INDEX TO ENTRIES

Note: This index is in a single alphabetical sequence, comprising personal names, place names and subjects (where the last two do not form the first word of the heading of an entry, e.g. Abortion Law Reform Association, Edinburgh Ladies' Educational Association). Personal titles and titles of periodicals and books are given in italics.